TELe-Health

Series Editors
Fabio Capello
Giovanni Rinaldi
Giovanna Gatti

For further volumes:
http://www.springer.com/series/11892

Fabio Capello • Andrea E. Naimoli
Giuseppe Pili
Editors

Telemedicine for Children's Health

Editors
Fabio Capello, MD, MSc
Pediatrics and Child Malnutrition
CUAMM – Doctors with Africa
Padova
Italy

Giuseppe Pili
Department of Child and Adolescent
Psychiatry and Neurological Disorders
ASL 1 Imperiese
Sanremo (IM)
Italy

Andrea E. Naimoli
Tech Department
Airpim Inc.
Wilmington, DE
USA

ISSN 2198-6037 ISSN 2198-6045 (eBook)
ISBN 978-3-319-06488-8 ISBN 978-3-319-06489-5 (eBook)
DOI 10.1007/978-3-319-06489-5
Springer Cham Heidelberg New York Dordrecht London

Library of Congress Control Number: 2014944559

Printed on acid-free paper

Springer is part of Springer Science+Business Media (www.springer.com)

Foreword

Telemedicine is the use of medical information exchanged from one site to another via electronic communications to improve a patient's clinical health status. This includes a growing variety of applications and services using two-way video, email, smartphones, wireless tools and other forms of telecommunication technology. Telemedicine lends itself to several clinical applications and is of growing interest in developed and developing nations worldwide. Starting out over 40 years ago with demonstrations of hospitals extending care to patients in remote areas, the use of telemedicine has spread rapidly and is now becoming integrated into the ongoing operations of hospitals, specialty departments, home health agencies, private physician offices as well as consumer's homes and workplaces. Research is limited on the clinical usage of telemedicine in pediatrics. However, it may have important implications for accessing pediatric subspecialty services, determining future health care workforce requirements and their distribution, improving communications with parents of sick and chronically ill children, and extending the boundaries of the medical home.

Telemedicine for Children's Health is a timely and comprehensive book illustrating the many clinical applications of telemedicine and showing its potential to facilitate the delivery of education and teaching programs as well as the facilitation of administrative meetings of health services to rural areas.

The three authors, Capello, Naimoli, and Pili, are well-seasoned experts of the field and the depth of their knowledge is clearly evident in the reported experience gained in years of clinical and research activity. The result is a very readable and useful handbook for all healthcare providers who treat children, as it covers different aspects of this complex matter. There is a number of logistical factors which are important when considering the development of a telemedicine service including site selection, clinician empowerment, telemedicine management, technological requirements, user training, telemedicine evaluation, and information sharing through publication.

The text is divided into four main chapters and it is easily accessible with clear organization, case presentations, and clinical advice highlighted in specific key points. So, the way these numbers are presented are useful to clinicians counseling patients and their parents.

The book starts by reviewing the numerous fields of applications of telemedicine (Part I): for example, parents are able to send data and requests and to receive advice and indications, healthy behaviors can be promoted, and children at risk are more easily signalled; furthermore, it is possible to contact parents to help and properly assist a child with a chronic condition as well as to inform parents in relation of their children's state of health or acute medical condition during school hours. Pediatricians can also exchange medical data to continuously monitor life-signs or other medical parameters for a home-based therapy both for acute or chronic conditions as well as ask for clinical information related to previous consultations.

The second topic (Part II) is related to complex technical issues of telemedicine. Indeed, a range of communication technologies are used – including email, telephone correspondence and videoconferencing. However, a layer of complexity can be created to offer different solutions for different users, without compromising the nature of the communication or of the contents. In addition, major issues can be raised, including cost-effectiveness, power supply, and privacy concerns.

A specific section (Part III) is then dedicated to the potentially great impact of telemedicine in rural and extreme rural settings and in developing countries. As a matter of fact, for people living in the rural areas the distance to main metropolitan centers often places restrictions on access to essential services, including specialist healthcare. Telemedicine provides a possible answer. Many different terms such as telehealth, telecare, online health and e-health have been used but they all have a common meaning, i.e. the use of information and communication technologies to deliver health care services at a distance.

The final part of the book discusses the relevant issue of distant learning (eLearning) for health professionals that work with children, basically aimed to teach healthy lifestyles for a high quality of life in young subjects.

The text utilizes nice images and clear tables to illustrate or to highlight key points and major concepts. The references are pertinent and insightful throughout the text and provide maximum value. Throughout the text, key points are highlighted and summarized in boxes. Most chapters also end with the authors' recommendations, a description of the chapter author's interpretation. The book is very fast and pleasant reading. This text would be a valuable reference for healthcare providers interested in telemedicine and clinical management of children. In addition, this book would also be of interest to medical students and residents, and sophisticated patients and other lay persons who want to learn more about this matter.

In conclusion, this volume is well worth its price and is highly recommended for both general and specialist pediatricians, including those beginning their careers and to hone their skills. It provides a practical approach to the recognition and management of several aspects of this complex matter, which is not a separate medical specialty. Products and services related to telemedicine are often part of a larger investment by healthcare institutions in either information technology or the delivery of clinical care. Provided that a reasonable number of conditions require a direct medical intervention, the introduction of telemedicine system and device for the delivery of primary care for children's health should be considered as a part of the process of improvement of the quality of care for children.

Genoa, Italy Pasquale Striano, MD, PhD

Preface

The increased demand for health, together with the availability of new and very advanced technologies, is today opening new scenarios in research and in the delivery of care. The use of communication devices is already offering new solutions aimed at solving the needs of the population, reducing the burden, and consequently improving the quality of life.

Childhood is a critical phase. Children are supposed to play out their existence under the sun, explore their creativity, and exploit the joyful nature of their age. Instead, acute or chronic medical conditions, together with social or psychological discomforts, can deeply affect the growth of a child, and therefore the lives of their parents or guardians. In an evolving world in which the role of children is changing, these are not unusual occurrences: the stress that young people are exposed to is high in both low- and high-income settings.

Children from industrialized countries have to live their lives in new frantic setups, whilst kids from the least developed regions have to fight the ancestral battle for food and against illness. The backgrounds are different, but the aims are not: easing the load related to a disease, and promoting healthy lifestyles are a shared goal, especially for children.

This is where telemedicine – intended as delivery of medical aid from a distant source in terms of prophylaxis, prevention, health education, diagnosis, prognosis, therapy, rehabilitation, and follow-up – can play a decisive role.

Improving the quality of the medical service, enhancing the possibility of making an early or difficult diagnosis, or providing expert consultation to find an optimal solution, and therefore a better therapy, is possible. Creating a convergence of professional competences coming from different areas in spite of the physical location of the consultant offering the counseling is an intelligible argument. Reducing the number of days of hospitalization for a child, and therefore the collateral sufferance of a disease, is another.

However, telemedicine, especially for children can offer even more.

Children have the right to live a quiet and joyful life. This is the case in every part of the world. Yet, today children worldwide still die of preventable or easily curable diseases. This should not be considered tolerable, as long as every child on this planet should have the same opportunities in terms of health and well-being.

On the other hand, tele-pediatrics cannot be considered as merely another form of medical consultation that makes use of the communication technologies and that takes place between patient and doctor, or between doctor and doctor. Indeed, all the actors and stakeholders involved in the delivery of care are included.

Parents, relatives, schoolmates, teacher, educators, are all actors in the management of a child's existence, and potentially promoters of good health, i.e., health intended as well-being and not only as a mere absence of illnesses.

What is more, though more: children have to become the main actors of this process, and consequently children should be the ones for whom these new tools are intended.

This implies that children must play a central role. Tele-medicine devices are not only for professionals; they can actually offer a means of creating connections among different health workers, but the real revolution that ICT technologies can bring to pediatric care lies in the fact that the youngest children are today able to understand and know how to use them.

In other words, these systems have to be built so that children can operate them and therefore play an active role in the management of their state of health.

Yet, there are still many challenges that developers have to face: telemedicine means covering distances. But what distances are we talking about? This is not a trivial matter, as such distances could vary from a very short range (neighborhood) up to a very long one (continents). However, of the two actors involved, the one that is asking for help, and the one that is delivering care or consultation, such distances cannot be covered just through communication technologies. It is true that we can perform a video-call from one place to its antipode with very little investment nowadays. But what about the beliefs, the ethics, the will to share, the way of life, and the way of thinking of the two people that are interacting with each other? What about the legal implications in terms of accountability and liability? What about the economic impact and burden? Above all, what about the real well-being of the child who is asking for care?

This means that distance work cannot be carried out if the role and the needs of the child involved in the care are not properly considered. Nevertheless, the use of telemedicine tools – when applied to pediatrics – has to first and foremost improve the quality of life of the child. Even if they cannot be ignored, any other concerns should be dependent on that consideration. Unfortunately, even in telemedicine that may not be the case.

It is a wild world, indeed, in which the logic of the e-business runs the risk of overcoming children's real needs and consequently their rights. Besides, it is not only a matter of economic interests, but rather the application of models of R&D that are not appropriate for children's health. It often happens when commercial processes that are suitable for enterprises are applied to the very complex world of human health. This would open the old Pandora's box of eHealth and its application, but this is not a coincidence: tele-medicine is a major component of electronic health.

When it comes to children, though, some diatribes should be put aside. This is not rhetoric, but common sense: children are the future, and there would be no house if the foundations were rotten. In this way the leading role cannot be played by parents, doctors or policy-makers, but rather the those who require health; in other words, the children.

And it is around children that every medical system, whether it is based on distant or traditional care, should have to be created. Thus, to the complexity of the human being, the complexity of a child's nature must be added, a nature that is also made of invisible things and of magic, and that could change over the time, place, and over the different stages the child goes through as it grows.

An accurate analysis of the request for help in today's multifaceted and multicultural world, of the role of children in the different cultures, and of the real needs in terms of children's health and well-being – also taking into consideration different age groups – is therefore crucial.

However, that assessment of needs has to start from the child, always considering what should be the best for him/her, taken as a single individual and as a member of a family and of society.

It also becomes clear that many other sides of the story have to been told: as the latest document on telemedicine from the European Union [1] strongly underlines, the legal aspects of telemedicine have to be considered, as long as different laws and different levels of liability are in use in different nations. Unfortunately, modern medicine cannot separate legal implications from everyday practice, although in the last few decades this led to the development of a defensive medicine, aimed more at protecting doctors and caregivers from persecution, rather than the needs of the patients.

On the other hand, we cannot shrug off rules and regulations, as they are the fundamentals of modern societies, even if laws have to cope with the society's needs and not vice versa.

Besides, the correct and accurate use of the resources that have to be invested in people's health – and not wasted – has to be taken into account.

Nevertheless, we have to start from people's needs or – in this case – from those of the children. Laws, regulations, and economic matters must therefore follow, creating a mask that could help stakeholders and policy-makers to take the right decisions, in an attempt at offering equal opportunities to everyone [2].

Therefore, the purpose of this book is not to offer given solutions, starting from the available technologies or regulations. Nor is it the goal to give a quick overview of the possible application of ICT devices and models for pediatrics, or to analyze the economic and the legal burden that could undermine the development of telehealth today.

Instead, we would like to introduce those clear and present needs in children's health that telemedicine is asked to address and possibly solve in the near or distant future. The only way, in fact, to plan working models is to assess the real needs in pediatrics today, trying to understand the context in which those technologies must be applied.

The planning and the engineering of ICT devices and software is the next step that planners and developers are asked to not undertake unless they have clearly in mind what is useful and what is applicable in the different scenarios for which telepediatrics is intended.

A starting point, for everyone, should be the child itself and the magic of that stage of life. Childhood, in fact, "is the time for children to be in school and at play, to grow strong and confident with the love and encouragement of their family and an extended community of caring adults. It is a precious time in which children should live free from fear, safe from violence and protected from abuse and exploitation. As such, childhood means much more than just the space between birth and the attainment of adulthood. It refers to the state and condition of a child's life, to the quality of those years [3]."

Padova, Italy — Fabio Capello
Wilmington, DE, USA — Andrea E. Naimoli
Sanremo (IM), Italy — Giuseppe Pili

References

1. European Commission (2012) Communication from the commission to the European Parliament, the council, the European economic and social committee and the committee of the regions. eHealth Action Plan 2012–2020—Innovative healthcare for the 21st century. COM(2012) 736 final
2. Angood PB (2004) Telemedicine, the internet, and world wide web: overview, current status, and relevance to surgeons. World J Surg 25(11):1449–1457
3. UNICEF definition of childhood. Available also from: http://www.unicef.org/sowc05/english/childhooddefined.html

Contents

1 Perceived Needs in Pediatrics and Children's Health: Overview and Background 1
Fabio Capello, Andrea E. Naimoli, and Giuseppe Pili

Part I Fields of Application

2 The Community 15
Fabio Capello

3 Telemedicine in Acute Settings and Secondary Care: The Hospital 37
Fabio Capello and Giuseppe Pili

4 Management at Home: The Chronic Child 51
Fabio Capello and Giuseppe Pili

5 Overtaking the Distances: The Child with Special Needs 61
Giuseppe Pili

Part II Technical Issues

6 Connectivity, Devices, and Interfaces: Worldwide Interconnections 71
Andrea E. Naimoli

7 Technology and Social Web: Social Worldwide Interactions 79
Andrea E. Naimoli

Part III Complex Scenarios and Special Settings

8 Rural and Extreme Rural Settings: Reducing Distances and Managing Extreme Scenarios 91
Fabio Capello

9 Telepediatrics in Developing Countries: A Better Care for Children in Low-Income Settings 99
Fabio Capello

Part IV e-Learning

10 eLearning: Distant Learning for Health Professionals That Work with Children 111
Fabio Capello and Andrea E. Naimoli

11 Health eDucation: Teaching Healthy Lifestyles for a High Quality of Life 117
Fabio Capello

12 Conclusions 123
Fabio Capello, Andrea E. Naimoli, and Giuseppe Pili

1 Perceived Needs in Pediatrics and Children's Health: Overview and Background

Fabio Capello, Andrea E. Naimoli, and Giuseppe Pili

Since the first draft of the Declaration of the Rights of the Child in 1927 and its approval in 1959 in Geneva and its following revision in 1989 in New York [1], many steps have been taken in the recognition of the role of children and toward the improvement of their conditions worldwide. The new conception of well-being also when referred to underage people has become more and more relevant in the planning of strategies for care.

The definition of the needs of the youngest and the consequent implementation of plans for the promotion of their health are crucial points in the policies of governments of industrialized and developing countries. Those same efforts are strongly supported by government and nongovernment organizations also in least developed areas.

The same idea of health indeed cannot be considered an universal value if the needs of the weakest ones are not taken into account. Children are the frailest part of our society, somehow, but they have the most important value at the same time. They are the tomorrow's people, and any policies for the future have to start from this principle, aside from any other goal they are aimed for.

Yet, is it really childhood an universal value to protect? And if it is the case, from which different standpoint is this value observed among different cultures? The answer is not trivial: the new vision of the family in Western countries sometimes collides with the old-fashioned role of the child that can still be observed in

F. Capello, MD, MSc (✉)
Pediatrics and Child Malnutrition, CUAMM – Doctors with Africa,
Via S. Francesco, Padova, Italy
e-mail: info@fabiocapello.net

A.E. Naimoli
Tech Department, Airpim Inc., Wilmington, DE, USA
e-mail: andrea.naimoli@elementica.com

G. Pili
Department of Child and Adolescent Psychiatry and Neurological Disorders,
ASL 1 Imperiese, Consultorio di Sanremo, Sanremo (IM), Italy
e-mail: giuspili@gmail.com

F. Capello et al. (eds.), *Telemedicine for Children's Health*, TELe-Health,
DOI 10.1007/978-3-319-06489-5_1,

more conservative settings. Children are undoubtedly a richness and a burden (mostly a joyful one) at the same time. Children are an investment for the future. Children are what fulfill people's lives.

On the other hand, the richness children bring to a family can be merely related to their role as a working force or to a prosecution of the cultural and physical inheritance of a tribe (a continuity over time and space of a family or a race). There is no right or wrong in those visions, but it comes to itself that the idea of well-being especially for children varies largely among different cultures.

On the other hand, why are we so constantly interested in children's health, and why do we still consider a crime toward a child as the most brutal offense? This is a transversal value that is shared all around the world, no matter how poor or technological is the background people come from.

Yet, are children the beating heart of our society, the representation of innocence, or the working force of the future?

Telemedicine has to connect people from close to very long distances. But are those distances really fillable? And can technology be the means able to fill them?

Diversities have to be considered and distances have to be overtaken.

1.1 Filling the Distances

What a child is and what is considered to be best in order to achieve the better possible quality of life change a lot according to the place the child comes from. Whereas people that belong to high-income or very high-income countries tend to consider childhood as a reservoir in which the child has to grow, to learn, and to become a member of the society, in other parts of the world, children are considered expendable goods for the family.

In Western countries, children are constantly protected and their wellness has a higher priority in the family affairs. On the other side, in very low-income areas and according to the traditional idea of the child, boys and girls are a burden for the families, but also a source of income. Children are hungry mouths and a family's link among past and future at the same time. In most rural and very poor settings, neither a man nor a woman's life can be considered complete unless a child arrives. On the other side, in a significant number of cultures, child well-being is secondary to the one of adults, in the same way airplane safety rules ask grown-up people to put on their oxygen mask in case of need before helping children to wear theirs.

Adults have to stay fit and well, in order to produce those incomes that help the family to go on. Besides elderlies – especially in a culture with oral traditions – are the memory of the race and the repository of wisdom. Old people have conquered their role in the society and are a precious point of reference for the youngest, while children are still not.

Nonetheless, this is only a schematic view of those scenarios, as long as those features are generally blurred or only partially expressed. Those same features indeed can be universally observed, in spite of the level of development or wealth of a country or of a culture.

Table 1.1 Some of the current differences in the vision of the family and of the child in different settings

Industrialized countries	Developing countries
Low number of children per family	High number of children per family
Modification of the vision of the family	Conservative vision of the family
Stable couples (can change over the time)	Many partners (even when marriages are consistent)
Children live their whole childhood with parents	Children can live with other next of kin
Children as the central and more vulnerable part of the society	Children as expendable members of the society
Children are cared for only by parents and close relatives	Children and families live mainly outdoor
Children and families live mainly indoor	Children are cared for by the whole community
Multiethnicity (as a resource and as a possible source of discrimination)	Mono-ethnicity (up to tribal clashes in very low-income settings)
High-resource settings	Low-resource settings
Educational programs	Illiteracy and lack of educational programs (including health education)
Mothers have a limited number of pregnancies in their fertile lives	Mothers have several pregnancies in their fertile lives
Mothers have to care for a limited number of children at the time	Mothers have to care for a great number of children at the time
Breastfeeding and weaning time do not depend on resources and/or on a new pregnancy of the mother	Breastfeeding and weaning time depend on the availability of resource and/or on a new pregnancy of the mother

Some basic differences among the two extreme standpoints can be summarized in Table 1.1.

On the other hand, children's health is a crucial factor that cannot be misjudged. Children are the core of the society because they represent the continuity between old and new generations. It means that they have to grow well, learn well, and become productive members of the world they live in, not for the world itself, but mainly for their own sake.

This is true in every culture and it is a starting point on which a working model can be built upon.

It is not accidental in fact that six out of eight points of the Millennium Declaration and Development Goals [2] are related to children. Children's well-being is one of the most important indicators to assess the level of health of a country. The death rate below the age of five, for instance, can be considered a significant parameter to identify whether the most fragile members of the society are protected or not [3], the level of mother-child care being another [4]. Most of the people in developing countries die in a young age because of illnesses that in other settings can be easily diagnosed and cured. There is thus a strict correlation with the level of health of the nation and the state of health of its children.

In poor-income settings, the number of young people that live in a society is generally higher than the number of the elder ones. In addition, a poor environment is most prone to develop a lower level of hygiene, while the level of health education, a measure that could help to prevent some of these conditions, is poor. It means that the risk to contract an illness is increased above all in younger ages [5, 6].

Moreover, the problems of the poorest countries are the bound to become the ones of the richest ones. The demographic changes that took place in the last centuries clearly show it, both because of the importation of foreign diseases (from travelers or immigrated people) and because of the genetic recombination among different races [7–9].

1.2 Youngest People, Healthcare, and Settings

Children from developing countries are not the only ones that need continued attention. Young kids are major users of the national health systems also in industrialized countries. In fact, the ones that more frequently – directly or indirectly – ask and seek for healthcare are the youngest. It implies that a great part of the resource that can be invested in health has to be for the children.

On the other hand, children are most frequently prone to develop acute conditions that generally lead to poor or absent sequelae, if they are promptly and properly treated. Besides, some other conditions are self-limiting and do not require medical intervention at all. It means that most diseases in children (whether they can be treated with a modest or a major intervention) are likely to be cured for good, once the medical condition is over. Elderly people on the contrary are likely to develop chronic and complex conditions that need a constant and lifelong care. The more elderly people survive, the more they need assistance. Because of that, the logic behind the interventions in young or in grown-up people is completely different.

The level of intervention differs a lot according to the setting in which we are operating. Where medical condition is more related to communicable disease and where the number of young people is more than the elder ones (poor-income setting, developing countries), many small interventions[1], but for a very large amount of young individuals, are required.

Where the threats for young people are less, acute illness mainly refers to a considerable number of minor and non-life-threatening diseases (i.e., upper respiratory infections, gastroenteritis, urinary tract infection). In those settings, the level of care is generally higher with an increased availability of resources that allow the diagnosis of complex medical conditions. The incidence and the prevalence of chronic diseases diagnosed in childhood consequently increase. Consequently, health systems have to shift their interest toward other targets as the management of chronic conditions in children, the improvement of the quality of life in the youngest, the promotion of healthy lifestyles, and the role of prophylaxis and prevention.

The natural discrepancy that also leads to a difficult governance and optimization of the resources in different settings is clear. This is not only an ethical issue, but

[1] For life-threatening events as complicated malaria: a condition easy to diagnose and treat with cheap medication with a fast recovery and often poor long-lasting complications.

rather a matter of convenience. As we have seen, the problems of the least developed countries are due to affect the richest ones as well.

As we will see, this is where telemedicine can play a prominent role.

On the other hand, telemedicine cannot be considered merely the provision of care from a place in which the level of health standards is higher to another where the resources are limited; nor the passive delivery of knowledge from a given source to a less privileged receiver. This is part of what telemedicine is asked to do, but probably not the most important aspect. On the contrary, telemedicine should be a mutual exchange of experience and knowledge. It gives the possibility to access healthcare no matter what place you are from, reducing the burden for the child, increasing its odds against the illness, and trying to make its life as better as possible. In other words, it would allow the child to receive prompt and high-level assistance in acute or life-threatening events; receive medical consultation, even if the child lives in the middle of nowhere; reduce the time spent in the medical office or in the hospital; and receive more quality time to spend in those activities children are supposed to do, and it would reduce the ones who needed to seek or receive medical treatment.

1.3 Why Telemedicine for Children?

Provided that a reasonable number of conditions require a direct medical intervention, the introduction of the telemedicine system and device for the delivery of primary care for children's health should be considered as a part of the process of improvement of the quality of care for children. As we will see, the empowerment of the child in the management of its own health (and therefore of its quality of life) is mainly a process of communication strategies that help the patient (or the kid) to understand what state of health he or she has and how he or she can communicate and express his or her own discomfort to caregivers or other people.

Communication also means a continuous feedback from the counterpart, in order to establish a closed circuit in which problems and solution could be continuously exposed.

Two main approaches consequently have to be considered:

- The use of telemedicine device to send information related to the child's health to health centers
- The use of ICT tools to gather health information, promote healthy lifestyles, prevent dangerous behaviors, or identify diseases in their early stage

These two approaches can or cannot include the active role of the child. In the latest case, an open communication can exist between the doctor (or the primary care center) and the next of kin of the child: a mother can ask for medical consultation using a tablet's app that can help to assess the health condition of her child, sending information to the surgery department or to the hospital nearby. A teacher can inform the health staff in relation to a child's psychological discomfort through an intranet Web 2.0-like connection and receive feedback and indications without leaving the teaching room. A general practitioner from a rural area can send a life parameter to an emergency department or to a poisoning call center and receive a specialist consultation live in videoconference. A researcher can acquire important health information that comes from the devices that children and teenagers use in their everyday life [10].

Yet those are merely a power-up of already existing solutions that can definitely improve the delivery of care and the role of prevention, but that still do not consider the central role of the child and above all do not exploit the potentialities of telemedicine, when applied to children's health.

On the other side, the same idea of primary care includes the active intervention of the health institution in the society. Because of the peculiar nature of the child, and because of its peculiar role in the society, primary care cannot be considered only as a relationship between parents and doctors. All the institutions that work in behalf of the child have to take part to the process. It includes primarily the school, where children spend most of their daytime. This can be particularly true when we consider the psychological life of the child: primary care and prevention have to deal also with the complexity of the life of a child. Bullying is, for instance, an important example of the hard experiences a child has to go through while coming to age. But even if teachers and educators are perfectly aware of the difficulties a child has to face in his everyday life, they still observe the problem from an external standpoint.

In addition, the same ICT tools that can promote health can be a source of problem that could have a terrible impact on a child's health. An impressive number of depressive conditions and of attempts of suicide in the tween or teen ages are referred to be related to computer- or Internet-mediated harassments [11, 12].

This is why telemedicine has to be for children: firstly, it can help children to have an active role on the management of their own health, which is supposed to result in an improvement of the quality of life both as children and as future adults; secondly, telemedicine systems can help children to establish proper communication channels with health providers that can help to send and receive information in a way the child can easily understand.

1.4 The Child-Doctor Encounter

Children live in a world that is different from the one of the adults. Furthermore, every child lives in its own world, in which the rules of the grown-ups do not apply. There is a lot to be discovered in such a place that is both inside and outside the mind of the kid. The problem is: is the door to that world open to everyone? To understand how medicine can help children, the first step is to understand what their expectations, their aims, and their desires are and the goals they aim for. But that is not easy; children have their own vision and their own communication strategies that can conflict with the one of the adults – places where dragon speaks and princesses have to be saved, where stuffed animals eat their cereals, and where a magic spell can make dad come home from job earlier.

Understanding a youngster could be difficult. Nevertheless, to reach what children keep inside, adults have to listen to what they have to say.

That has to be done according to the little ones' rules.

There cannot be a medicine for children in fact (and therefore a telemedicine, which strictly depends on that) without a proper assessment of a child's necessities. What can appear as a mere exercise is indeed the crucial phase of the management of children's health.

It all starts from an everyday pediatrician-child consultation.

A common encounter generally starts with a next of kin of the child entering the consultation room – whether it is a GP surgery, a community midwife's office, or a triage bench of an emergency department – carrying alongside the child or the children he or she is seeking help for. The main rule of engagement says that doctors or health workers are supposed to address directly the child in order to offer a suitable environment, reducing the stress and allowing the child to fully express itself. Yet most of the conversation lies between the health worker (the doctor, the nurse clerking the patient, the triage officer collecting the medical history) and the parent or the next of kin who is seeking for assistance. Children are often excluded from the conversation, and the collection of symptoms is often a doctor-parent affair.

There are two main methodological errors in that. First of all, the child is left outside from the game. It is not the attention he deserved that is discussed, but what the parent feels about the child's health. Some conditions, as minor injuries are, do not need in the abstract, for the child intervention. This is a logical error as long as the child's discomfort is the first issue that medicine is asked to address. But subjective feelings cannot be submitted by third parties, even when an illness is suspected. The first methodological error, therefore, is the exclusion of the child's standpoint from the medical-patient discussion.

The second point is the level of conversation that is generally brought on the high side. Doctors tend to interrogate parents above toilet behaviors, Fahrenheit or Celsius degrees of fever, long-lasting expectoration, or strangely smelling secretions. Most of the words used are over the head of the children, and many of the discussions are impossible to be grasped by the kids.

It brings a third consequence: as long as most of the dialogue is between the parent and the doctor, the age of the child became a secondary matter.

The first step for the design of a model of care, and above all of a model of distant care, is to understand of the role of the child and of the way it can understand its state of health and the way it can communicate it to others.

To overcome those methodological errors underlined, it is crucial to start from the last question that arose, proceeding backward:

Who is the child we are caring for?

Is he or she a newborn, a toddler, a kindergarten kid, a schoolboy or a schoolgirl, a tween, a kid in early teens, or a boy or a girl?

The second step is how the doctor, the nurse, or the health worker that is addressing the patient has to establish contact with the kid. Most of the consultations, for instance, are conducted with a desk between the doctor and the patients – even more often with a computer screen parting them – or with a doctor that has a higher standpoint compared with the one of the child (because the doctor is standing in front of him or her while he or she stands or sits, because the child is laying on a bed or on a couch, and because the two of them sit on chairs of the same height). It creates a sudden disparity that undermines the possible relationship that could occur between the child and the doctor (Fig. 1.1).

But communication is a complex thing that involved different parties and that could be put in danger whenever the message cannot be correctly delivered. There are different reasons for communication and different theories that explain how messages can be distorted by environmental noises or wrong interpretations.

The basic point, yet, is the delivery of the message. And no message can be delivered if it is not properly produced. It means, in other words, that the doctor that aims to work with children – or better for the children – has to learn how to communicate with them. It implies that he has to use the right words, but it also means that he has to use the right channels and to press the right switch that could allow the kid to receive the messages and communicate or to make him or her receptive and eager to reach the message or to establish a relationship.

If the encounter is with a male kid, for instance, in order to establish a communication channel, the doctor or the health worker could talk about the last football match, possibly using the same argument to create examples that could help him to express his discomfort (which is another way to collect history and to clerk the patient) or in a later stage to explain the disease and the consequent treatment. In the same way, Pixar's last motion picture or a princess' tale from far far away could help to open the door with preschool children. Or asking about the last teen idol on the block could open the way with tween girls. There are no rules in that, but the most important thing is to use words and examples kids could understand and could be longing for.

The last point is to understand the role of the child during the consultation. It is not possible that the whole gathering of information passes through the parent, even when the child is big enough to clearly express itself. Parents are keen to give an abundance of symptoms and to stress those ones that they think are more important. But not necessarily those signs or symptoms are lived in the same way by the kids.

Aside from personal changeability and sensitivity, the same pain – evaluated, for instance, with a pain scale, like the six-grade smiley faces [13] – could be felt differently among children and adults and among children of different ages or medical condition [14, 15].

That does not mean that children have to be the only actors that interact with health professionals: first of all, children do not or cannot ask for help alone. They refer primarily to those figures that live inside the family or inside their closest circle, even when they are suffering a lot. Second, asking children to take care of their own health alone would be a big responsibility that not every child could or should have to manage and that could apply only at particular age groups.

On the other hand, help-seeking behaviors have to be evaluated with great attention: children cannot ask for help by themselves in some cases, but can show their distress in numerous ways that do not include direct talking. And those children that require

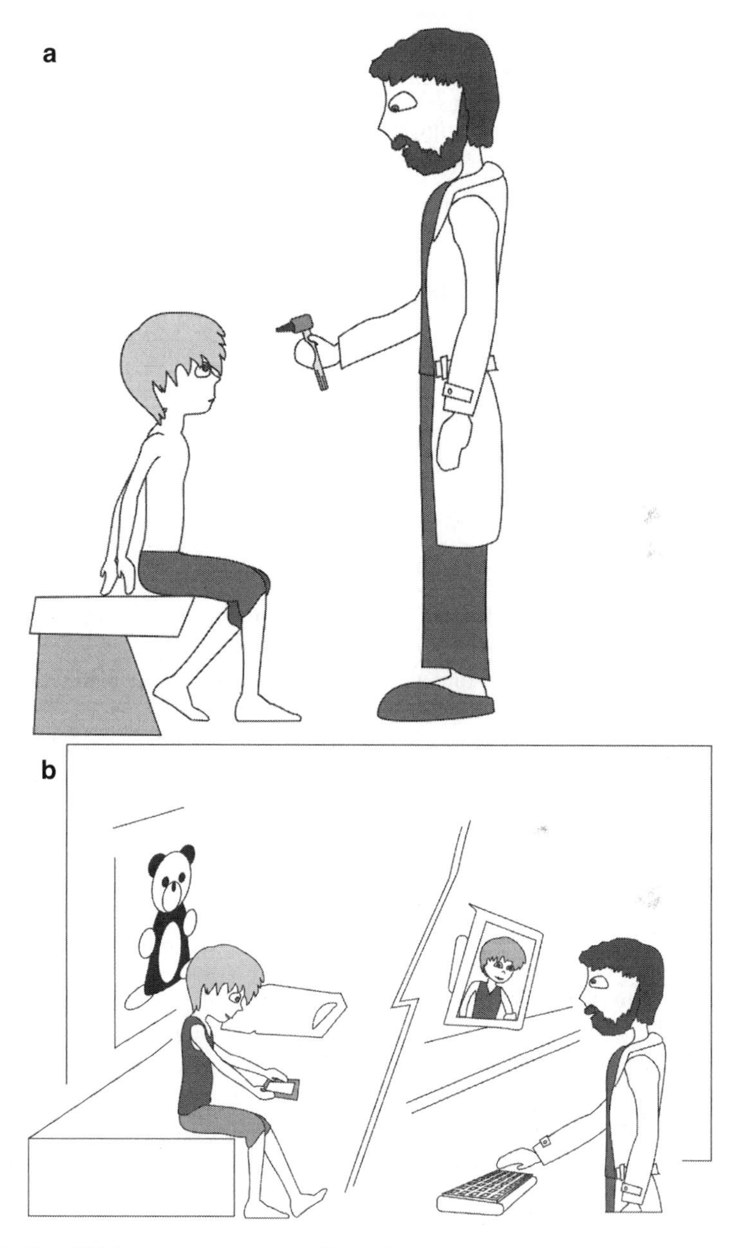

Fig. 1.1 The child-doctor encounter generally produces two standpoints: the one of the child that is usually in a lower position and the one of the doctor that has a dominant position. This produces a disparity between the two subjects, the child also being naturally weaker (in terms of knowledge and cultural baggage, physical strength, or emotional control) when compared to adults. Telemedicine devices can subvert the points of view, leveling the level of communication or inverting the leading positions. It can help especially in older children to open a channel of communication with the health professional. The intrinsic risk is that electronic communication could undermine the interaction, excluding all those sensory perceptions that are a fundamental part of the communication itself, increasing the distance between the child and the doctor

more attention could possibly have a problem that is deeper than the suspected organic condition they could suffer of. Also that has to be studied and evaluated.

Besides, the same nature of consultation changes accordingly to the way the child could enter in contact with the health worker and through which channel it reaches the point of care. It varies a lot depending on which health system is delivering the service and if a family pediatrician is accessible to the community. Many health systems though have community doctors or community nurses that can deliver the care right into the natural environment of the child, offering the possibility to schedule an appointment in a very short time or to give consultation without previous advices [16].

1.5 Which Technology for Whom?

As we have seen so far, the definition of standards for the development of telemedicine for children has to consider which goals these new tools are intended for. On the other side, the design of specific devices has many threats that have to be avoided. Three main approaches have to be considered:

- The introduction of new devices according to an enterprise model that do not take into account the peculiarity of healthcare, but follow economic models already in use for the development of new communication technologies. Those are bound to fail because the management of health is complex at least as the communication is, as the experience with electronic health records had already shown [17, 18].
- The design of specific systems able to deal specifically with the complexity of the world of children's health. It implies the assessment of the real needs and the development of models that consider all the problems related to health and the way they interact among themselves (including social, deontological, economic, legal, ethical, geographical, personal issues), keeping in mind that every single person is an individual and not the sum of its average value. The commercial model is apt to offer solutions that consider mostly the economic standpoint and possibly the assessment of the perceived needs. As long as medicine is not a profitable enterprise, the ideas of efficacy, productivity, and cost-effectiveness are different from the ones of business systems. It means that specific devices have to be designed against the economic models and therefore are bound to be expensive, while their implementation can take place only in specific or advanced points of care (if telemedicine is considered only as a mere teleconsultation system, those health centers, because of its nature, would not need such complicated communication devices).
- Using existing technology, already available, also designed according to enterprise-based models. Designing a device for telemedicine has to consider the economic impact of those same systems. Above all, if a standard is introduced only in a very small area, the possibilities that it can be used in very long-range communication systems (which is a big part of telemedicine) become small. As we will see in the second part of the book, the introduction of devices has to

consider the opportunity that those devices could be spread in the community. It made no sense to produce a very well-designed tool that requires a complex managing system and that requires to be universally adopted starting from zero. On the contrary, the use of already available nets and equipment could ease the introduction of new technologies and applications. Touch screens, for instance, have revolutionized electronic devices: whereas once in the designing and the production phase all the possible functions of a device had to be designed, today the interface can change and can be modified according to the new algorithms that can be programmed in the device (whether there are firmware updates or new apps specifically designed for a specific aim). The limit of a new device today is its computing capability (that could be on the other side enhanced thanks to the use of the net or of grid-like systems) or the number and the kind of sensors it has. Besides, plug-in devices can expand the input and output possibilities of a device, so that the combination of data received from the combination of sensors it has and the processing of those same data are virtually countless.

Conclusions

As we have briefly seen, the design of new tools has to consider the final users and the way they can communicate with other users or with automatized systems. As long as the children's health is considered, children have to be the main actors of the communication and therefore of the telemedicine model. It implies that those devices have to be kids friendly first of all. But it also means that families and professionals have to be able to use them as well, with or without the involvement of the child.

Lastly, delicate matters have to be dealt with due care. Children have to be the master of their own health, but many are the issues that they are not able to understand. The use of Internet-based devices could expand the possibilities to a boundless scenario. Yet they can offer access to a threatening world that stands over the head of a child. Misunderstanding and misinterpretation of medical conditions are already a major issue today: people use the Internet to try to understand medical problems that affect them or the ones they care for, ignoring or heedless of the unreliability of the information found on the Web. Children are even most exposed to this threat, as long as they do not yet have the power and the knowledge to distinguish what is possibly wrong and what is possibly true.

References

1. Declaration of the rights of the child. Adopted by UN General Assembly Resolution 1386 (XIV) of 10 December 1959. Available from: https://www.un.org/cyberschoolbus/humanrights/resources/child.asp. Accessed on 6 Mar 2014
2. UN General Assembly. United Nations Millennium Declaration. Resolution 55/2. 8th plenary meeting, New York 8 Sept 2000
3. Guillot M, Gerland P, Pelletier F, Saabneh A (2012) Child mortality estimation: a global overview of infant and child mortality age patterns in light of new empirical data. PLoS Med 9:e1001299

4. Målqvist M (2011) Neonatal mortality: an invisible and marginalised trauma. Glob Health Action 16:4
5. World Population trends. United Nations Population Division. Department of Economic and Social Affairs (DESA). Available from: http://www.un.org/popin/wdtrends.htm. Accessed on 5 Mar 2014
6. Catalogue of Population Division Publications, Databases and Software. Reports on world population trends and policies and on population estimates and projections. UN Department of Economic and Social Affairs Population Division. Available from: http://www.un.org/esa/population/pubsarchive/catalogue/catrpt1.htm. Accessed on 5 Mar 2014
7. Austerlitz F, Heyer E (1999) Impact of demographic distribution and population growth rate on haplotypic diversity linked to a disease gene and their consequences for the estimation of recombination rate: example of a French Canadian population. Genet Epidemiol 16(1):2–14
8. Herbinger KH, Drerup L, Alberer M, Nothdurft HD, Sonnenburg F, Löscher T (2012) Spectrum of imported infectious diseases among children and adolescents returning from the tropics and subtropics. J Travel Med 19(3):150–157. doi:10.1111/j.1708-8305.2011.00589.x, Epub 2012 Feb 24
9. Gushulak BD, MacPherson DW (2004) Globalization of infectious diseases: the impact of migration. Clin Infect Dis 38(12):1742–1748
10. Linder LA, Ameringer S, Erickson J, Macpherson CF, Stegenga K, Linder W (2013) Using an iPad in research with children and adolescents. J Spec Pediatr Nurs 18(2):158–164
11. Bucchianeri MM, Eisenberg ME, Wall MM, Piran N, Neumark-Sztainer D (2013) Multiple types of harassment: Associations with emotional well-being and unhealthy behaviors in adolescents. J Adolesc Health. 54(6):724–9.
12. Daine K, Hawton K, Singaravelu V, Stewart A, Simkin S, Montgomery P (2013) The power of the web: a systematic review of studies of the influence of the internet on self-harm and suicide in young people. PLoS One 8(10):e77555
13. Bosenberg A, Thomas J, Lopez T, Kokinsky E, Larsson LE (2003) Validation of a six-graded faces scale for evaluation of postoperative pain in children. Paediatr Anaesth 13(8):708–713
14. Campos MJ, Fraga MR, Raposo NR, Ferreira AP, Vitral RW (2013) Assessment of pain experience in adults and children after bracket bonding and initial archwire insertion. Dental Press J Orthod 18(5):32–37
15. Logan DE, Williams SE, Carullo VP, Claar RL, Bruehl S, Berde CB (2013) Children and adolescents with complex regional pain syndrome: more psychologically distressed than other children in pain? Pain Res Manag 18(2):87–93
16. Weiner JP (2012) Doctor-patient communication in the e-health era. Isr J Health Policy Res 1:33
17. Rinaldi G, Gaddi AV, Capello F (2013) Medical data, information economy and federative networks: the concepts underlying the comprehensive electronic clinical record framework. Nova Science Publishers Inc, Hauppauge. ISBN 978-1-62257-845-0
18. Capello F, Gaddi A, Manca M. Ehealth, care and quality of life. Springer, Milan. 2014 ISBN 8847052521

Part I

Fields of Application

The Community

2

Fabio Capello

2.1 Telemedicine and Primary Care: Fields of Application

In order to understand which ICT tools can be appropriate in the development of a telemedicine system for children's health, the different approaches to care have to be considered. There is a basic one that is provided by the families and that follows the child also in the further steps of the care. There are primary care and community-based care, aimed to give a continuous support and early intervention in the same environment the child lives in. There is secondary care, a second level of intervention where specialist consultations can take place and complex or acute conditions can be managed (hospitals and specialist clinics). There is a third level of care that focuses on follow-up, long-term treatments, and rehabilitation.

A community-based care depends mainly on primary care (general practitioners, community nurses and midwives, community health and social workers) and on home-based treatments (families). The same primary care can offer a range of solutions and interventions that have to be considered and upon which a telehealth system can be built.

To understand the implications of a community-based care, it is important to underline what the idea of well-being in the youngsters is. Children's health can be defined as "the extent to which individual children or groups of children are able or enabled to (a) develop and realize their potential, (b) satisfy their needs, and (c) develop the capacities that allow them to interact successfully with their biological, physical, and social environments [1]."

It is clear that it is not only the idea of disease and treatment that we have to deal with. Thus, pediatricians and community workers delivering primary care for children are called to respond to a number of issues that go beyond the mere definition of illness. Health in fact does not mean absence of disease anymore [2].

F. Capello, MD, MSc
Pediatrics and Child Malnutrition, CUAMM – Doctors with Africa,
Via S. Francesco, Padova, Italy
e-mail: info@fabiocapello.net

F. Capello et al. (eds.), *Telemedicine for Children's Health*, TELe-Health,
DOI 10.1007/978-3-319-06489-5_2,

Nowadays, some of the criticalities children's doctors have to deal with include exposure to adverse environmental conditions, language and cultural diversities, poverty, health insurance-related problems, and illegality (as in illegal immigrants' children, legal matters that include also privacy and parental responsibility – managing and being responsible of any communication related to the child's health such as results of tests, medication schedule, revision schedule, prescription renewal, update of the state of health of the child). The same use of medical equipment presents some problems: doctors in fact should "be fully conversant with any equipment" they use, as long as they have to "ensure that it has been properly serviced and is in working order before beginning any procedure" [3].

In addition, pediatricians are facing in their everyday activity "more complex conditions with roots in societal problems." There are many examples of such complex conditions. Asthma, food related problems and obesity, mental health problems, oral health problems, and violence are examples that require thinking past the clinical encounter. They are among the most important problems that affect children's health and are strongly influenced by factors in the family, the community, the society, and the environment. Pediatricians are now being trained "to take a population approach and urged to think beyond the walls of the clinic when addressing significant health issues [4]."

For a good medical practice, in fact, the doctor should "treat patients as individuals and respect their dignity. Work in partnership with patients: listen to, and respond to, their concerns and preferences; give patients the information they want or need in a way they can understand; respect patients' right to reach decisions with you about their treatment and care. Support patients in caring for themselves to improve and maintain their health [5]."

Considering all these aspects, a *patient- and family-centered care* is extremely important in pediatrics: the family and the environment the child lives in are the primary sources of strength and support. Besides, those provide and record important perspectives and information to the child that are vital in clinical decision-making [6]. In these contexts, the use of telemedicine can help to empower families and children, so that they can become directly involved in the management of the health of their child (*telemedicine offered to families, children, teachers, educators, volunteers in order to communicate with social and healthcare workers*). This is part of a more complex system, in which health professionals are connected to a higher level of consultation with other specialists able to offer consultations, advices, and supports (see Fig. 2.1).

The family doctor's office, on the other hand, remains the first place where most of the issues that are related but not strictly dependent to the child's medical condition can be discussed and identified. Some of those – as the emotional and behavioral concerns that go with a disease or with a particular state of health – are well known and are part of the disposition of the single child or are a normal component of the human nature. Some are the product of the modern societies as well as the legacy of old-fashioned issues (i.e., cyberbullying, child's addictive behaviors, child exploitation).

In addition, the GP is called to take crucial decisions for both acute and chronic conditions that spread from the delivery of a treatment to the management of

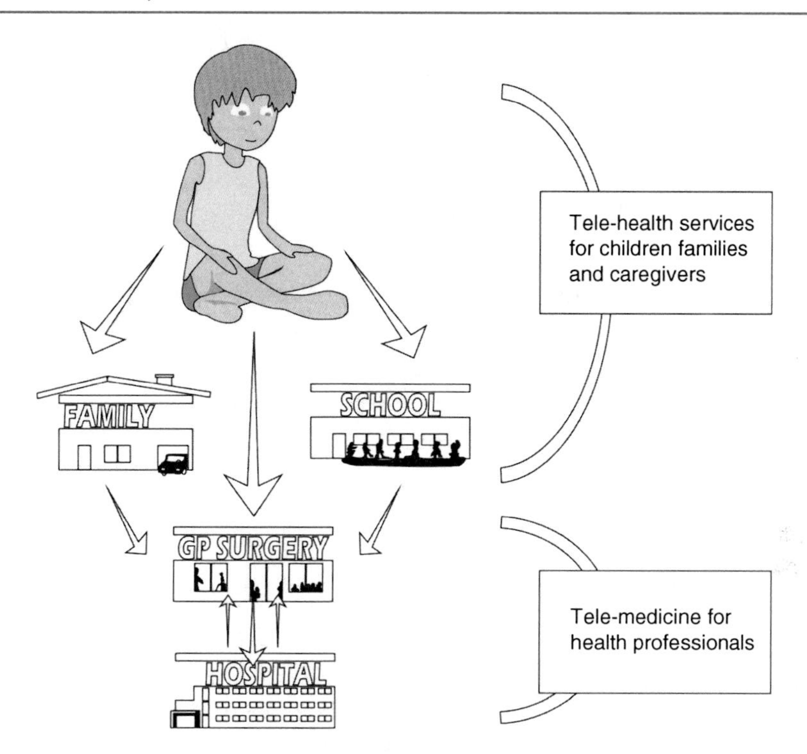

Fig. 2.1 Two different levels of care, connected in a unique model thanks to telemedicine systems. The child interacts with a net of care that communicates with the health and social professionals. Those same health providers can communicate at a professional level to other specialists, giving and receiving feedback from children, families, and the community (see also Fig. 2.3)

in-depth investigations that require laboratory or instrumental tests or specialist consultation. In that, they can be supported by a system for telecommunication and distant assistance (*telemedicine for GP or local pediatrician for consultation with specialists and advanced health centers*).

In an attempt to create a framework in which the telemedicine approach can be introduced, the main activities of a doctor that works in the community have to be disclosed and clarified: several are the areas of intervention as well as numerous are the fields of application of those services. That has to be planned and implemented according to the needs to be covered. The actions or fields of intervention a pediatrician working in the community or in primary care is asked to cover are summarized in Table 2.1.

In addition, childhood is not a static condition. Children evolve over the years: problems and medical conditions, as well as normal pattern of growth, are different according to their age and the gross age group they belong to (see next paragraph and Fig. 2.2). Because of that, those areas of intervention can be specific only for a particular phase of the growth or can go through the whole childhood, up to the teen years and the adult age.

It appears clear, analyzing those different fields of intervention, that diagnosis and treatment are only a component of the delivery of care in the community when

Table 2.1 The main areas of intervention that children's doctors are asked to cover, whether they work in the community- or in a hospital-based care

Monitoring children's physiological development
Assistance to parents
Safety at home and prevention of home accidents
Prophylaxis of common diseases or of special conditions
Primary, secondary, and tertiary care
Early intervention and for acute or emergency conditions
Early diagnosis and detection of the onset of medical conditions (both acute and chronic)
Assessment of an ill child
Diagnosis and investigation of acute and chronic conditions
Consultation among health professionals to assess a health condition and make a diagnosis
Monitoring of the therapeutic regimen both in acute or in chronic conditions
Monitoring of chronic conditions, follow-up during convalescence, early detection of relapses
Referral to secondary health center, specialist, and hospitals for further medical consultation and treatment
Follow-up
Rehabilitation
Monitoring of possible deviation of the growth curve
Support for parents with children with medical conditions
Health education
Prevention of early unhealthy behaviors (unwittingly fostered by parent caregivers)
Prevention of later unhealthy behaviors (fostered by the same child)
Promotion of healthy lifestyle
Prevention of addictions, substance abuse, and risky behaviors
Prevention and detection of child abuse (sexual, physical, and psychological abuse) and neglect
Prevention and detection of child harassment, bullying, and psychological discomfort at school or in other community activities

it refers to children's health. The lion's share is taken by several other issues that compose the main part of a child's life (such as behavioral habits and alimentary habits). That is because children are naturally healthy and most of the medical conditions that can affect them are acute and self-resolving. On the other hand, the lack of intervention and management of all these aspects can lead to health problems in the adult life. That is why, for instance, a particular attention is generally given to child alimentation. Child obesity is becoming a crucial issue in Western countries, where the onset of noncommunicable diseases on the adult age is strictly related to habit learned during childhood. This is why several studies to prevent, assess, or reduce obesity in children, thanks to the use of telemedicine tools, have been produced so far [7–10].

But if the aim is that children could learn healthy habits since their early years, that implies that the child (and the his or her family) has to be empowered so that he or she can have an active role and assume healthy lifestyle not just like an imposition, but rather as a normal and acceptable part of his or her life. Hamburgers, pizzas, and french fries are welcome (in small amount) in a child's diet as long as they are part of a most diversified nutritional regime. Yet a child is likely to ask for those

Fig. 2.2 A tailored system for telemedicine has to consider the child not as a static entity. Children change abruptly over the years and so do their interests or the way they see or interact with the world. It implies that (a) the specific needs of each age group have to properly considered and (b) the systems for telemedicine have to exploit the interests and the communication strategies that are appropriate for each age group

categories of food unless a richest menu is presented. It means health education but also the awakening of the idea that all the foods are good and all the foods can be tasty. But what role can children have if a telemedicine system designed to monitor their diet merely collects data on a child's food habits and transmit anthropometric parameters to a remote server? For sure, it can help a GP or a pediatrician to monitor a treatment, but it does not help the child to become aware of his own state of health or of the implication that an unhealthy behavior can have on his future adult life.

Thus, a telemedicine system has to be designed for children and families and according to the age of the child (as interests and communication strategies vary among the years; see Fig. 2.2).

It does not mean that such devices have to have cartoon characters or teen's idols painted on their shell. They may have and they should have them, but that is not the point. The key to get in touch with children is to use a communication strategy that they can understand and exploit. Older children, for instance, would be able to interact with a telemedicine system by themselves. But it implies that the same system has to be thought about (and therefore designed) in the same way children think. Children have to learn how to deal with those tools, being involved in their use.

They have to be guided (by parents, teacher, and doctors) especially in the early ages, so that they can become acquainted to and not scared of them. But in the late childhood and during the coming of age, they should be also accessible in private. Sometimes in fact older children might not be keen to talk of their health problems only with their mom or dad.

Besides, a telemedicine system has to deal with all or with part of these aspects of a child's health. But the perfect telemedicine system also has to be so flexible that a number of different actors could simultaneously interact with it. It means that it must have different, interchangeable, and properly designed point of access. How many? As many as the user's typologies are. Parents can be interested in the use of telemedicine tools to send data and request and receive advices and directives. Teachers should communicate with health and community workers, to promote healthy behaviors or to signal children at risk; or need to contact parents in order to help and properly assist a child with a chronic condition; to inform parents in relation of their children's state of health or acute medical condition that set on during the school hours. Sport team's coach could be interested in work in connection with pediatric specialists to promote healthy lifestyles (physical activities, diet) or to monitor the progresses after a post-injury rehabilitation process. A pediatrician can ask for medical data to continuously monitor life signs or other medical parameters for a home-based therapy for both acute and chronic conditions or ask for previous medical data and clinical information related to previous consultations (in a multidisciplinary approach or to review a medical file of a child that, for instance, once lived in a different town or area) or ask for referral or teleconsultations to other specialists, sending and receiving clinical data in real time.

2.2 Telemedicine in the Community

Primary care is generally intended for those people that (a) live in a limited area and need to access the basic medical services and (b) need continuous health services among the time (e.g., a pediatrician that follows a child from birth up to his puberty). According to this view, most of the telemedicine services are intended to connect the different actors of the society in which the child lives in: the child itself, the parents, the schoolteachers, the educators, those volunteers that work with children, the GPs, the nurses, and other medical health workers (at local levels, at hospital levels, or at higher levels). As we have seen, telemedicine can spread from next-door connections to intercontinental communications. In this case, the net is most likely intended to connect the local officers that from different standpoints deliver health services to children. The same criteria used for a local community can be extended virtually to the entire world. This approach can be of some help in the planning of long health services (according to a community-based model) in rural areas, in extreme scenarios, or in developing countries (see Part III).

We will see in Part II which are the main strategies that can be used in telemedicine to deliver health services and which one of those is most appropriate for children's health. The ways telemedicine can work especially in a closed environment as the community are:

- *Instant messaging and telecommunication services* (multimedia and videoconferences, Skype, VOIP, telephone, mobile phones, bleepers, and pagers): They substitute a conventional face-to-face consultation and work mainly in real time.
- *Text-based and multimedia-based services* (SMS, MMS, FB-like messages, instant chatting, open forum, communities, access to databases): Can deliver information mainly for the record that can or cannot require an immediate action or a feedback (e.g., the practitioners send text messages to their patients to inform about the opening hours of their surgery; a mother sends the value of glycemia of her diabetic child via email; a teenage child asks a community nurse for information about intimate matters). Those can also support a real-time consultation or can contribute to create knowledge and spread top-to-bottom and bottom-to-top information (communities, user-generated contents with moderators, wiki-like resources, virtual consensus conferences).
- *Device-based services*: Medical devices (also known as point-of-care devices [11]) that send in remote to doctors and specialists medical information collected at home or from other distant sources. They could send the report automatically (for instance, through a Wi-Fi connection), or the value expressed can be reported by the caregivers at home using secondary ICT communication devices (emails, SMS, instant messaging, specifically designed apps).

Most of those same services can today exploit technologies already in use, relatively cheap and largely available as small smartphone devices, tablets and apps, or personal computers.

In an ideal system, all the actors that take part in the communication systems in behalf of children's health (kids, parents and relatives, doctors, midwives, community nurses, psychologists, teachers, educators, coaches, social workers) should be able to access and share medical information, with a level of liability and confidentiality. This is part of a most complex scenario as the eHealth is, in which telemedicine is only a component of the whole. Yet an integrated system, to which different applications and devices could be connected to, can offer a whole approach.

The owner of the whole system, besides, had to be the child. It does not mean that a kid has to take care of himself or herself alone or that the ones that care for him or her have to be put aside. It should be a shared governance of the child's health, in which the child is the central figure of the model.

It means that the whole net has to consider the child as the main actor, able to communicate in an easy way with all the other actors that are interested in the improvement of his state of health and therefore of his quality of life, both in his young and in his consequent adult life. It has to be stressed that a child is not a static entity as the way children establish relations with the outer world depends strongly on age, sex, and personal or family background (Fig. 2.1).

Besides, in very early ages (newborns, toddlers), the parents have to take this role, them being an extension of the little one that connects the child (and its need) to the world.

In other words, children are not asked to take care of themselves alone, but to participate and be involved in the management of their own health. This is why telehealth systems have to be designed according to their level of comprehension and their will to communicate.

The strong point in a telemedicine system, in fact, is that the level of communication can be set according to the level of the child that as we have seen (and as we will see in the case of chronic or special conditions) is different according to the age, the personal profile, and the condition of the child. It implies that the barrier that blocks the face-to-face or traditional encounter can be overtaken. This can be considered as a Copernican revolution, in which the child becomes the center of attention.

For older children as tweens or teenagers, this can be a major issue: children of these ages are experiencing a rupture with their early childhood that is needed in order to create the adult they would become. This is, though, a complex and challenging process. Help-seeking behaviors are common, as the child longs to receive help and advices. But those supports cannot come from parents, as the child is not eager to share its internal life with them. Neither does it have to be, as the break with the standards of its family is the trigger of this process.

This indeed is a fundamental part of the development of a child's personality that expresses itself in a breach (a real crisis) with its childhood beliefs. Parents are not the person they aim to talk to, yet many questions – most of them related to health issues – have to be answered. Friends and the Internet are a common source of information and in some way of medical and sexual education, but many other most reliable and reassuring sources can coexist.

This somehow destructive (sometimes self-destructive) energy can be canalized into a productive process.

2.2.1 Telemedicine at Home (For Children and Parents)

There are several reasons a general practitioner or a community health worker can be sought for. Yet, when it comes to children, parents seem to be most sensible to some particular issues or medical conditions. Therefore, most of the contacts are related to a few cadres of interventions that depend mostly on the age of the child: asthma, allergies, respiratory infections, attention and learning disorders, abdominal pain, dermatological issues, toilet habits, prophylaxis, and prevention [12, 13]. Those requests for help vary according to the age of the child and the family's background [14].

For example, during the first years of life, most interventions are related to the physiological development of the child or to reassuring parents for mild or very mild conditions. On the other hand, in those years, some chronic or congenital conditions can come to the attention of the doctor.

Later in the life of the child, the pediatrician is asked to solve more acute or subacute conditions or to deeply investigate some chronic problems that still do not have a diagnosis. Schematically, the kind of interventions requested (in terms of assessment, management, and timing referral) is summarized in Table 2.2.

Table 2.2 Some of the most common reasons for a family child doctor visit during the different age groups

Age group	Physiological	Pathological	Psychological and behavioral
Newborns	Check of the physiological development Feeding-related counseling	Investigation of congenital diseases (malformations, syndromes, chromosomal aberrations, infectious diseases acquired during pregnancy)	Child care advices
Infants	Check of the physiological development Feeding-related counseling Weaning Vaccination schedule	Acute medical conditions (UTI, URTI, reflux, colic, gastroenteritis) Investigation of early-onset chronic conditions (mostly congenital)	Child care advices Milestone achievement
Toddler	Check of the physiological development	Acute (and chronic) medical conditions	Child care advices Toilet habits Autism disorder
Preschool age	Check of the physiological development	Acute medical conditions (URTI, abdominal pain) Deeper investigation and follow-up of chronic conditions	Learning disabilities Toilet habits Autism disorder
School age	Check of the physiological development	Acute medical conditions Deeper investigation and follow-up of chronic conditions Mental diseases Acquired chronic conditions	Enuresis and toilet habit Depression and anxiety Harassment and bullying Learning disabilities
Tween	Check of physiological and sexual development	Acute medical conditions Deeper investigation and follow-up of chronic conditions Mental diseases Acquired chronic conditions	Enuresis and toilet habit Depression and anxiety Harassment and bullying Learning disabilities Relationship abilities
Teen	Check of physiological and sexual development	Acute medical conditions Deeper investigation and follow-up of chronic conditions Mental diseases Acquired chronic conditions STD Early pregnancy	Depression and anxiety Harassment and bullying Learning disabilities Relationship abilities First sexual experience Eating disorders Addictions

According to this frame, four main areas of intervention can be considered:

(a) Monitoring of the physiological development
(b) Identification of acute or subacute medical conditions that require for immediate intervention (and possibly a specialist referral)
(c) Identification, investigation, and follow-up of chronic conditions that still do not have a diagnosis (and possibly a specialist referral)
(d) Psychological help and support diagnosis (and possibly a specialist referral [15])

Another fundamental area has to be added to the previous that will be discussed further in this book (see Chap. 5), related to the monitoring and the management at home of the child with a chronic disease.

In addition, the level of feedback in a home-based care can be related to:

(a) Automatic response
(b) Online or real-time response
(c) Delayed response

Those depend on the final aim of the application and the organization of the management of the child at home (e.g., (a) an automatic response that adverts the parents that the inserted value of glycemia is safe, and no other actions are required; (b) a real-time video consultation where the doctor visually assesses the child's conditions, gives advices, and reassures the family or suggests further actions if needed; (c) test results submitted via email that the doctor can examine when he can, sending a report to the families and suggesting further actions if needed).

Telemedicine has to consequently deal with these needs, and the development of ICT for the management of children's health has to start according to these considerations, taking into account the different aims that vary according to age and profiles and the scopes of applications and devices.

2.2.1.1 Monitoring of the Physiological Development from Birth to Teenage Years

As shown in the previous table, this aspect includes two different facets: the physiological aspect and the pathological one. Children do not come with a user's handbook, and often parents are keen to ask for medical intervention in order to solve many of the problems that are related to their role. It includes, for instance, how many times a baby has to be fed or what is the most appropriate time to change a nappy, give a bath, or go to sleep (physiological aspects). On the other hand, the early detection of acute or chronic disease in the early years of life is often strictly related to the monitoring of the growth of the child, as often a problem or a delay of the growth rate is correlated to an underlying disorder. Moreover, a parent will appeal to a doctor in case of fever or secondary to the onset of other acute symptoms (pathological aspect).

The first one is an easy field of intervention for telemedicine devices. Two main areas of intervention can be adopted:

(a) *Monitoring of the physiological conditions*: Devices that retrieve anthropometric information (as automatic scales connected to the Internet) and automatically send this information to a remote server. Those devices are easily accessible and are based on easy and already available technologies (computer, smartphone,

Internet connection, and so on). They can be used for routine checks in infants or older children to assess the normal growth, or they can be used for specific needs.

(i) *Basic monitoring of the child's growth*: In this first case, those devices can substitute the normal medical-patient encounter, reducing the stress of a physiological assessment in a medical environment for the child. As we have seen, that can be connected to eHealth systems that can plot the information in an electronic health record (EHR), where those data can be consequently accessed, processed, and stored.

(ii) *Provide support and follow-up related to the physiological development*: In this case, a specifically designed device or software can monitor specific parameters. An example could be the test weighing before/after nursing to detect the amount of milk sucked by a baby from the mother's breast. In this case, a normal scale can be used, but the device has to know that the information collected has a specific purpose (it has to store the weight before and after breast feeding, calculate the difference – which corresponds to the milk sucked by the baby – and give a feedback, also considering the amount of milk already taken previously by the child and the growth curve adjusted according to its age). The way parents set the new transmission mode therefore has to be simple and easy to access. On the other hand, those kinds of tools can augment the level of stress for parents and children: pediatricians, for instance, do not suggest the use of test weighing in children that do not have growth problems, as long as it produces more stress in the mother and augments the level of complexity of the moment of feeding, which on the contrary has to be a natural moment between a mother and her child. Integrated systems have to help parents understand whether a teleconsultation (or the use of telemedicine devices) is advisable or not.

(b) *Provide suggestions and information related to the physiological development*: This is a fundamental step that is strongly related to the development of healthy behaviors and lifestyle. Parents are constantly asking for help, especially in Western countries where families are generally smaller and the number of children or child's relatives is reduced. Many parents deal with children for the first time when they have their own baby. It creates a discomfort that can result in health problems for the child during his development. The possibility to access reliable information, and possibility to contact a health professional when he is needed, is a crucial point of telemedicine for children. Yet many parents still prefer the use of social networks, forum, or webpages to share their doubts and their fears with other parents (or strangers). That can indisputably lead to very dangerous behaviors as the advices that they can retrieve on the Internet are not reliable and generally do not come from health professionals. On the other hand, people sometimes need to listen in simple words from other people like them to feel that everything is fine or that everything is going to be alright. Web 2.0 has to be considered as a possible solution: it can offer an immediate, home-based assistance, and it can mediate the level of intervention, so that dangerous advices can be removed. It can give the needed assurance and level of confidence to parents, provided that the source of information is adequately screened. Other tools can be used to provide such a service: videoconference, app-based tools, and multimedia websites. As long as telemedicine is not a one-way intervention (parents seeking for information on the Web) but is intended to connect two distant poles, the redirection to webpages, forum, or multimedia contents should have to be mediated.

Also in this case, these considerations might apply to all ages with some differences. For example, a new mother could need advices on how to wean her child, whereas a father of a teenager will need to be advised about the correct dietary intake. A telemedicine system can plot the growth of the child and receive guided information related to the introduction of new foods in the child's diet. Besides, the ICT tools – especially when designed for those purposes – should have to be customizable and/or should adjust themselves according to the function and the specific user they are interacting with.

2.2.1.2 Identification of Acute or Subacute Medical Conditions That Require for Immediate Intervention

Many requests for help in the community still depend on acute conditions that require quick assessment, a quick diagnosis (whether they are upper respiratory infections, abdominal pains, or other minor issues or major medical conditions such as pneumonias, meningitis, or appendicitis), and in some cases a quick referral to a health unit or to a hospital or an advanced health center. This is a major concern for parents and for children. Thus, a priority for telemedicine has to be the management of all those acute scenarios, that cannot be delayed but rather need of almost real time interactions.

There are at least two major areas in which the use of ICT devices can help the family to assist their child in those circumstances. Those refer to the two main tools doctors still have to assess a patient, namely, anamnesis and examination. Theoretically, anamnesis can be easily collected with telemedicine systems (point A), whereas OE remains a more complicated matter, due to the fact that it traditionally implies a direct interaction between the doctor and the patient (point B). Telemedicine, therefore, has to substitute these two approaches to the disease with:

(a) *The use of direct consultation with standard ICT tools and videoconferences*: They can give a direct contact with the doctor or the health workers. The main problem and the history can be clarified while the child remains at home. It would help to reduce the burden of the disease itself; would reduce the need for consultation to A&E departments; reduce the time from the onset of the pathology to the one of the diagnosis; and possibly can produce early interventions that could help minimize the consequences and complications. On the other hand, giving the possibilities to the doctor to assess the patient directly from home can reduce the use of improper self-medications, meanwhile reassuring the parents on the conditions of their child. A Web form can be used to tick the most important points of anamnesis (self-assessment), while distant connection can offer access to patient records that could suggest in real time the previous medical history of the child.

(b) *The use of telediagnostic devices*: The distant consultation lacks yet of a fundamental tool. Even if doctors could be able to see the child and immediately understand the level of gravity of the patient they are seeing in teleconference, several information could be missed, so that a proper diagnosis and triage cannot be done. The use of specific devices able to monitor the life parameter and to give specific information can help to substitute the physical examination, meanwhile giving other information that generally needs deeper investigations. Some of those devices are still under development or already in use, as the telestethoscope. On the other hand, it implies that a family has to have at home several and possibly expensive electronic devices just in case they would need them in the future. These issues may be solved with the use of smartphones or tablets that already have a number of physical sensors and that in the future will be likely to have more. Those would give more possibilities that can be exploited to measure physiological or pathological parameters. Yet this will be a controversial issue that will be addressed as long as no telemedicine system could survive if a wide and easy accessible net cannot be available. There would be in fact no telemedicine if the two sides of the story (child and doctor) are not connected.

2.2.1.3 Investigation and Follow-Up of Chronic Conditions That Still Do Not Have a Diagnosis

The first step while trying to make a diagnosis is to understand the symptoms and the history and to use the tools of the clinical methodology to reach the goal. Is it true that most of the following steps depend on the GP or on the pediatrician that is trying to make the diagnosis (request for further investigations, as laboratory exams, imaging, or request for specialist consultations)? On the other hand, the suspect of a chronic disease starts from an issue that children and parents are asked to express. In a home-based care, in which telemedicine device and techniques are used, the main strategies are related to:

(a) *The use of Web 2.0 tools*: That can ease the approach to the medical service for families or can push the child himself or herself to seek for help when he needs it. Chronic disease can show themselves as major health issues, but they also bring minor discomforts that nonetheless reduce significantly the quality of life of the child or can put at stake the health state of the adult life. Teaching the children and the families to ask for help in a proper way, providing the proper information, and supporting them during the whole course of the disease (not only at the moment of the diagnosis) are a crucial point. The other is to help children and parents to understand that there is a problem when they face one. ICT community tools can provide the correct information, reduce the time spent in the pediatrician's surgery (or the one needed to reach it and to wait for it), and reduce the need for consultation in those cases that do not need medical intervention.

(b) *The use of a device that can send 24-h information*: In the first stage of the medical investigation, when it is clear that there is an actual problem that has to be addressed and solved, the use of telemedicine devices specifically designed to collect specific medical data can help to reach the goal. Some medical instruments can be already considered telemedical devices, as the ECG Holters, that record at home some medical data that could be analyzed afterward without the need for hospital admission. In the idea of telemedicine, yet, a real-time reading of the data that could also give an online feedback should be achieved.

(c) *The use of ICT tools for secondary care and specialist consultations*: Because chronic and complex conditions generally need a multidimensional approach where different professionals can offer their area of expertise to solve a problem, the use of telemedicine devices (videoconferences, instant messaging, email) can help to achieve that goal. Nowadays, it is common that a family has to move from a place to another in order to get specialist consultation anytime a GP sends a referral for deeper investigations. GPs on the other side have their challenges in scheduling the appointment with specialists and in receiving professional advices. Yet telemedicine can create a net of professionals that can offer their expertise to give real-time or deferred consultations. One easy application is the videoconference among the family (at home), the GP (in his office), and the specialist that speaks from a distant hospital. On the other hand, this can be considered merely an innovative solution for a traditional approach. Telemedicine as part of an eHealth system can offer more than that, giving access to a number of information (also retrieved from online medical records) that parents, doctors, and specialists can access and use in real time in order to solve a problem.

2.2.1.4 Psychological Help and Support

Children, and especially the older ones, live in a tough world. Many are the threats that they can meet during their development and the coming of age. Prevention of mental discomforts and psychological distress has to be achieved, but the traditional ways of communication, because of their nature and above all of the nature of children and teenagers, does not generally work. Adolescents and preadolescents are experimenting new communication strategies that are part of their growing process. They have to leave behind the beliefs of their childhood mainly made of given dogmas, or magic thoughts, they have been used to. This is the way the communication among older kids and parents is not often of any help. Moreover, children are not keen to openly express their feelings with strangers. In the teen years, the kids of one's own age are the mirror of oneself. Nonetheless, the risk of misleading behaviors and of wrong information is strictly related to this age and to this kind of communication.

Children, on the other side, are the ones that are more open to receive the innovation that technology brings to our everyday life, and because of their young age, they grow with electronic devices and tools that are part of their everyday life but that were not still invented when the adults were younger (digital native generation [16]). This way, children are the ones that can better understand the potentiality of telehealth devices. But, again, telemedicine has to think as children think, and consequently telemedicine devices have to be designed so that children can use them.

A further reason is that not only teenagers need psychological support: younger children can have complicated lives that could go back to their very early ages.

The strategies that children develop to build relationships start since the kindergarten years, for instance, and most relationship problems can be addressed to this stage. On the other hand, there are psychological stresses that could arise since the early life, as, for instance, the loss of a parent or of some other important figures. In addition – as it will be discussed in Chaps. 5 and 6 – chronic diseases and special conditions require for continuous psychological and medical support that have to be delivered in the most comfortable way for the child. It implies that a home-based support, whenever it is possible, has to be preferred.

The same applies for parents that in many cases do not know how to deal with the complex world of a teenager or of a preteen and in many cases do not know how to relate with them or who can give them help and support in order to understand and help their kids.

Some of the possible strategies can make use of:

(a) Web 2.0 tools: Teenagers love to acquaint with kids of their age. This is the most simple and direct way to communicate. On the other hand, there would be no difference between a normal research for information on the Internet and telemedicine tools. Web-based communities, mediated by psychologists and health professionals, could offer a possible solution, also for the early detection of Internet abuse, harassment, and exploitation.
(b) Direct consultation with standard ICT tools and videoconferences.
(c) ICT tools to improve the level of consultation among different specialists (e.g., pediatrician-psychologist; teacher-pediatrician).

Those tools have to be appealing and do not necessarily have to mimic the traditional way of intervention that the child may distrust. They have to create a stable connection and a soft landing zone, where the child could feel confident to express its feeling, knowing that on the other side, there is really someone eager to give advice and support and above all that really cares.

It is crucial to underline that the child does not have to be let alone. Parents, teachers, educators, and caregivers have to be involved in the whole process. It is the precise duty of the health professionals to assess the children's needs and to help them to express themselves, with or without the direct involvement of the parents. Those communication channels, based on telehealth systems, have to be presented by doctors and educators, allowing the child to freely express himself but keeping always in mind that there must be a direct connection with the family. The same family has to back the child every time, even when his or her privacy and confidentiality (which always have to be granted) are safeguarded.

2.2.2 Telemedicine at School and in Social Places for Children

Most of the issues analyzed in the previous paragraphs can be stressed to their extents in a complex environment as the school is. Parents are particularly able to detect distress in their children, yet kids spend most of the time of their days at school, and many of the dynamics that happen in their young life happen during the teaching hours. For this reason, teachers and educators can play a fundamental role in the assessment of a child's state of health and in the early detection of psychological and physical problems.

But how can telemedicine help those professionals to improve the quality of life of the children they care about?

Two main strategies have to be analyzed: first the use of screening devices that could send data to remote operators in order to early detect pathologies and problems. In screening programs, where a high number of patients have to be screened in order to detect few cases, this is crucial. It makes no sense, for instance, to install a device that could detect the level of glycemia in the blood of a child in every home, as long as the possibility that a child could suffer of diabetes is relatively low. The school is a natural place where such screening could take place, also because it can overcome the problem related to the owning of expensive or complicated devices that could limit the diffusion of telemedicine at home. On the other hand, a device like that could have a role in a school attended by several schoolchildren: it can be used routinely on many children with the aim to early detect a possible fatal disease.

The second approach aims to give teachers and educators the possibility to enter in touch with health professionals in order to receive information and advice for those children that they suspect being affected by disease, medical conditions, or psychological problems and for those children that have a chronic condition and that require continuous medical assistance.

Teachers and educators can retrieve medical data and information related to a particular child's health (this requires a properly designed electronic health record or EHR); can ask for help from GPs and family doctors in relation to specific conditions or to signal or investigate a suspected illness; get in touch with parents and relatives in a continuous feedback system aimed to early detect any possible health problem; send medical information detected at school for those children that have a chronic condition or are still under medical investigation for a still undiagnosed disease; and get support in case of suspected abuse, neglect, bullying, and behavioral problems and in the prevention of addiction and substance abuse, depressive moods and suicidal thoughts, and psychopathological problems.

These approaches need a properly designed system able to connect the school with the doctor's office and the child's family and home: ICT tools can already offer possible solutions for a net like that, but a working model requires the knowledge of the communication strategies, especially when applied to children, and of the eHealth's theories. The risk, otherwise, is only to create a cage that surrounds the child.

All those systems have to respect the confidentiality and the privacy of the child; therefore, the access to delicate data that could undermine the right of privacy of the child or that are likely to expose the child to future damages or harassment should have to be carefully governed.

There is a further indication for the use of telehealth systems in social places for children like schools. Those locations can be the natural place where health education programs can take place. We will briefly analyze this aspect in the last part of this book (see Chap. 12), underlining the potentiality of eHealth and ICT tools to deliver eLearning and distant education.

2.2.3 Telemedicine at the GP Surgery and for Community Health Workers

Many of the challenges that GP and community health workers have to face today are related to a number of requests and conditions parents directly ask for. This is a new issue as long as the care for the child, especially in high-income countries, is dramatically changing. The request for health is increasing significantly: parents ask for a better health for their children, and as already said, health does not only mean the absence of diseases. The problem is that the idea for health for children is changing. Whereas in the past children were an expendable part of the society, today they are considered the weakest and most valuable figures. Once they had a productive role and represent a continuity, but the main actor in a family was the *pater familias* (with many differences among the different cultures). The rate of children's death was high, and families used to have many children in order to augment the probabilities for their offspring to reach the adult age. This is not different to what happens in many low-income settings today (see also Chaps. 2 and 10).

Today, children are considered the weakest part of society and the ones that require most protection. Thus, the request for health is increasing. Any minor condition requires a prompt diagnosis and solution. Most of the conditions that are commonly bearable and self-treated in adults (as grippes, flus, cough, or minor aches) are a major cause of medical intervention in children. Fever, for instance, is often addressed by parents as a very dangerous symptom even if there are plenty of very mild diseases that show themselves with a high increment of the body temperature [17, 18]. GPs and health community workers have in this new scenario a peculiar role. They need to solve very essential issues that have a top priority according to the family expectancies, but that can be considered minor pediatric problems. On the other hand, complex conditions have to be identified and properly investigated. The role of primary care is to detect among all the possible situations (the multitude of presentations among the many parents ask to solve) which one requires minor interventions and which one needs deeper analysis and assistance. The risk is to give a higher priority to minor conditions (e.g., a physiological reflux or a tummy ache) and to ignore insidious and potentially lethal pathologies (e.g., celiac disease).

The main challenge for telemedicine is to offer support to community health workers and GPs in order to help them to navigate in this deceitful scenario. The role of distant consultation has to be discussed and considered. It can be a mere substitution of a telephone call among different professionals and specialists. But it can also be an integrated eHealth system that can send and receive information according to specific technologies. That solution can range from a referral model that can help to optimize the flux of consultations among primary and secondary care to an automatic system that can give medical advice according to the presented scenario or support the medical decision. The discussion on the use of diagnosis support system is still open, and many are the issues still unsolved. Yet the future of telemedicine – especially as a support of primary care centers – is still strongly related to the use of automatic systems. As we will see in the next chapter, in fact, of the three main approaches (direct consultation that substitutes face-to-face consultation; call center-like consultation; automatic response) that can be used for instant messaging, automatic systems appear to be the less reliable but possibly are the most rapid solutions. The discussion is still open.

The role of community pediatricians and midwives on the other hand is to give advice to improve the quality of health and therefore of life of the children in the society. It definitely includes the role of prophylaxis and prevention. The widespread diffusion of ICT devices (as smartphone and tablet) can help to reach this goal: Vaccination campaigns, for instance, can exploit telemedicine services thanks to the use of automatic reminders (e.g., SMS or emails); screening programs can send reminders and feedbacks to improve the compliance and satisfy the request for health of the parents (push information or specifically designed apps); the outbreak of seasonal illness can be monitored with automatic systems [19] (e.g., GPS localization of the ICT device that belongs to the child – children that are at home during the school hours can be possibly be ill; that could be used in epidemiological screening or can alert a GP if the families gave authorization to access such personal data).

Finally, telemedicine can help to open a channel of communication among GP surgeries and hospitals, improving the quality of work for both (reducing useless referral or improving prompt diagnosis of specific conditions). Many of the referrals from the GP to the emergency department or the specialist health center are related to minor conditions that general practitioners are not able to solve in their office. Whereas specialists can easily classify a medical condition and establish with or without further investigation whether that condition requires hospitalization and advanced treatment, GPs – because of their generalist knowledge – are generally unable to manage such cases. It produces a number of referrals and hospital consultations that augment the health expenditure and increase the burden for children and families. Telemedicine is the quickest solution to this problem [20]. Many minor conditions should be investigated directly in the GP office, avoiding the need for a direct consultation in a hospital or secondary health center.

Telemedicine specifically designed tools can be placed in GP surgeries as well as traditional communication tools. As long as GPs deal with a number of patients, the use of telediagnostic tools, and possibly expensive tools, can be justified. As we have seen, families cannot afford the costs of very expensive devices that can or cannot be used if they live with a healthy child with no particular medical conditions. Families cannot justify the purchasing of a tele-ECG device, for instance, that has to be used just in case a child possibly develops a suspected heart condition. GPs on the other hand can have such a tool in their office. According to this vision, families can experience a specialist-level consultation in a next-door surgery.

That is likely to reduce the burden for the families and society (e.g., the number of work hours lost, money spent for transportation, pollution related to the use of private cars, the number of useless consultation in a secondary health center, clogging the facility and delaying other needful consultations).

Conclusions

A natural place for a kid is the family and the community he or she lives in. Any measure engaged in behalf of children's health at home and in their own environment is due to improve the quality of the action undertaken: delivering the care in a space the child is familiar with would reduce the burden and increase the odds of success.

Besides, school and after-school activities play a central role in the life of a child, and the involvement in those activities increase with the age group of the child. A constant evaluation could be possible to early detect discomfort in children, help to manage chronic conditions that need of a constant care, or offer health education programs.

Social workers, family doctors, community pediatrician, and health workers have a key role in primary care, but they cannot work alone. They instead need to connect with the families in a continuous process of evaluation, care, and feedback and to reach secondary points of care (hospitals, advanced specialist health centers) to send and receive information able to improve the quality of their work, to support them in complex decisions, and for quick and adequate referrals.

All those actors that contribute to the development of the child should be connected among themselves in a dynamic net able to find problems and offer solutions.

The use of telemedicine devices can help to achieve this goal, reducing as well the number of hours or days spent by the child at the doctor's surgery or in the hospital, giving back quality time that kids can spend with their families and with their friends.

This is due to reduce the burden and to improve the quality of life of the child, provided that those systems are properly designed, so that the child's needs (based on its age, background, and personal profile) are taken into account. Therefore, it is paramount that those technologies have to be built around the child, in a patient-centered care, and are designed so that he or she can easily understand and – with the support of the families and of caregivers – interact with them (Fig. 2.3).

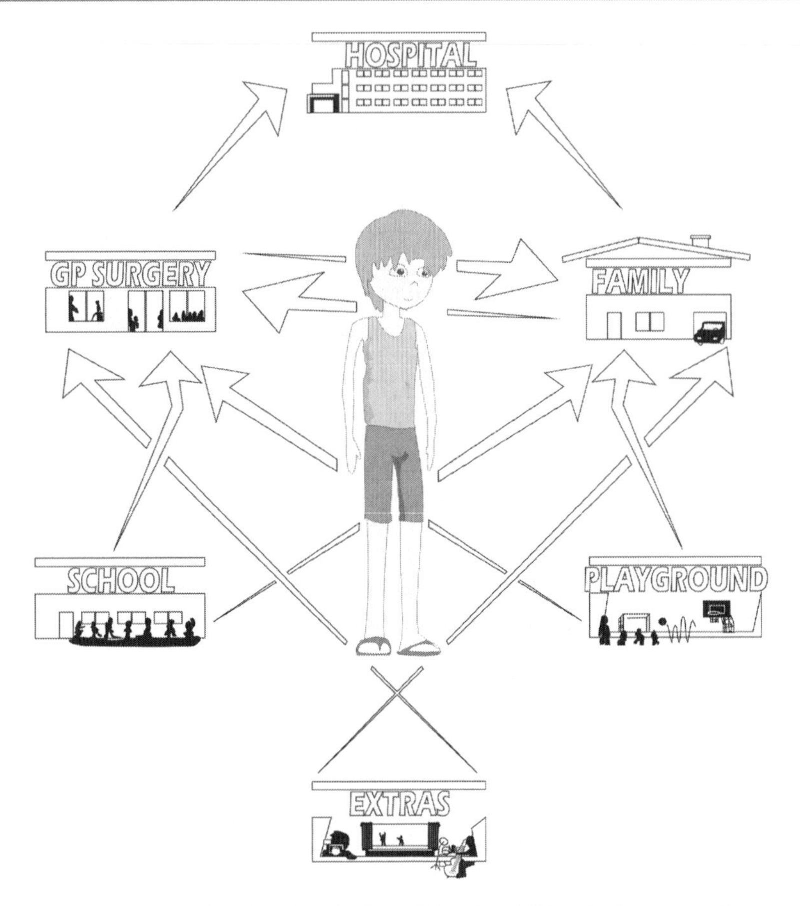

Fig. 2.3 The concept behind the idea of telemedicine for children in the community: a net of information connects the child and the institutions that take part in its development and care. The child has a central role and virtually can interact with the whole flux of information. If a health concern arises or when an intervention is needed, the whole net can be involved, under the guidance of the family or of the health professionals at different levels. ICT tools – both for generalist use (smartphone, tablet, PCs) and based on advanced and specialist devices – communicate among themselves, accessing when needed to a properly designed electronic health record (EHR) in order to input, read, and process medical and personal data

References

1. National Research Council and Institute of Medicine (2004) Children's Health, the Nation's wealth: assessing and improving child health. National Academies Press, Washington, DC
2. World Health Organization (1946) WHO definition of health. Official Records of the World Health Organization, no. 2. New York, p 100. Amended on 1948
3. MPS – The Medical Protection Society Common Problems. Managing the risks in general practice. MPs0645: 08/12 (last review, Aug 2013). Available from: http://www.medicalprotection.org/mps-uk-common-problems-gp-setting.pdf. Accessed 7 Mar 2014

4. Kuo AA, Etzel RA, Chilton LA, Watson C, Gorski PA (2012) Primary care pediatrics and public health. Meeting the needs of today's children. Am J Public Health 102(12):e17–e23
5. GMC – General Medical Council (2013) Good medical practice. General Medical Council, Manchester. ISBN 978-0-901458-60-5
6. Kronick JB, Hilliard RI, Ward GK (2009) The changing role of the paediatrician in the 21st century. Paediatr Child Health 14(5):277–278
7. Toftemo I, Glavin K, Lagerløv P (2013) Parents' views and experiences when their preschool child is identified as overweight: a qualitative study in primary care. Fam Pract. 30(6):719–723
8. Davis AM, Gallagher K, Taylor M, Canter K, Gillette MD, Wambach K, Nelson EL (2013) An in-home intervention to improve nutrition, physical activity, and knowledge among low-income teen mothers and their children: results from a pilot study. J Dev Behav Pediatr 34(8):609–615
9. Lipana LS, Bindal D, Nettiksimmons J, Shaikh U (2013) Telemedicine and face-to-face care for pediatric obesity. Telemed J E Health 19(10):806–808
10. Davis AM, Sampilo M, Gallagher KS, Landrum Y, Malone B (2013) Treating rural pediatric obesity through telemedicine: outcomes from a small randomized controlled trial. J Pediatr Psychol 38(9):932–943
11. Musaad SM, Herd G (2013) Point-of-care testing governance in New Zealand: a national framework. N Z Med J 126(1383):72–79
12. Data and statistics. CDC – Center for Disease Control and Prevention. Available from: www.cdc.gov. Accessed 7 Mar 2014
13. Data Resource Center for Child & Adolescents. The child and adolescent health measurement initiative. 2012. Available from: http://cshcndata.org. Accessed 7 Mar 2014
14. Data From the National Health Interview. Summary health statistics for U.S. children: National health interview. Vital and Health Statistics, U.S. Department of Health & Human Services – Centers for Disease Control and Prevention. Hyattsville. Series 10, No. 258. Survey, 2012
15. McLennan JD, Sheehan D (2008) Where do young children in speciality care come from?: http://www.ncbi.nlm.nih.gov/pmc/articles/PMC2247443/. a preliminary investigation of the role of primary care physicians. J Can Acad Child Adolesc Psychiatry 17:20–25
16. Oliver Joy. What does it mean to be a digital native? CNN. 8 Dec 2012. Available from: http://edition.cnn.com/2012/12/04/business/digital-native-prensky. Accessed 7 Mar 2014
17. Enarson MC, Ali S, Vandermeer B, Wright RB, Klassen TP, Spiers JA (2012) Beliefs and expectations of Canadian parents who bring febrile children for medical care. Pediatrics 130(4):e905–e912
18. Walsh A, Edwards H, Fraser J (2007) Influences on parents' fever management: beliefs, experiences and information sources. J Clin Nurs 16(12):2331–2340
19. Alex "Sandy" Pentland (2013) The data-driven society. Sci Am 309(4):78–83
20. Herendeen N, Deshpande P (2014) Telemedicine and the patient-centered medical home. Pediatr Ann 43(2):e28–e32

3 Telemedicine in Acute Settings and Secondary Care: The Hospital

Fabio Capello and Giuseppe Pili

A policy for the implementation of a telemedicine system has to strongly consider the role of the hospital as a main center for the management of children's health. It is true indeed that most of pediatrics can be practiced in the community and that the improvement of the children's quality of life starts at school and at home. Yet children are some of the main users of emergency facilities [1] both in high- and low-income countries. In addition, although children are more keen to develop acute conditions that generally require minor intervention that can be solved in the GP surgery, the chronic diseases of the young age are insidious and require for deep and complex investigations. Those same investigations generally require the intervention of a net of specialists – not different from the one of the adults – as long as general pediatrics cover only a minor part of all the possible conditions that could need an intervention in an ill child.

The complexity is augmented when we consider that not every hospital and emergency department has a specific pediatric ward or facilities and equipment to handle underage people [2].

Especially for minor events, family doctors and community pediatricians should be the first ones to see an ill child. Nevertheless, parents are keen to take children to the hospital, expecting to receive better care, even if surveys show how most conditions would not need to be attended by a medical doctor and should be managed in the community or at home [3, 4].

In any case, hospitals and advanced health centers play a central role in the delivery of healthcare to children.

F. Capello, MD, MSc (✉)
Pediatrics and Child Malnutrition, CUAMM – Doctors with Africa,
Via S. Francesco, Padova, Italy
e-mail: info@fabiocapello.net

G. Pili
Department of Child and Adolescent Psychiatry and Neurological Disorders,
ASL 1 Imperiese, Consultorio di Sanremo, Sanremo (IM), Italy
e-mail: giuspili@gmail.com

F. Capello et al. (eds.), *Telemedicine for Children's Health*, TELe-Health,
DOI 10.1007/978-3-319-06489-5_3, © Springer International Publishing Switzerland 2014

Table 3.1 The major area of intervention in children's health and the place where the care should naturally take place

What	Where	Who
Primary and secondary prevention, minor health conditions	Community and primary care centers	GP surgeries and community health centers
Diagnosis and treatment of more complex health conditions and special treatment	Secondary care centers	Hospitals and specialist clinics
Management of chronic conditions or disabilities	Tertiary care centers	GP surgeries and community health centers, clinics for long-stay patients

They cover three main areas that therefore have to be explored (see also Table 3.1), in order to develop telehealth systems able to respond to the specific needs of each one of them:

- The management of emergencies and acute conditions
- The management of specialist consultations, needed to make a complex diagnosis, and the role of specialist treatment
- The management and follow-up of chronic diseases (Table 3.1)

3.1 Emergency Settings and Acute Onsets

Acute medical conditions in children are a major cause of access to an advanced medical service in the early years of life [3]. This happens because children are normally fit and do not suffer from all the chronic conditions that reveal themselves in the adult life or that are secondary to additive conditions, sequelae and complications, relapses of previous diseases, accidents, and degenerative or late-onset disease. Chronic disease in children is generally due to a congenital or hereditary problem (including multifactorial diseases as the type I diabetes) or results of a previous untreated condition, of a very severe medical problem (e.g., complication of an infective disease as meningitis), or of an accident or trauma.

Less often families complain of divergences from the physiological processes, as a retard of the growth rate or of the accomplishment of the growth milestones or of minor issues that do not require a fast investigation (e.g., an ingravescent skin condition as impetigo, a changing nevus, or a progressing mycosis).

It implies that many parents ask for help only when the child is acutely ill, ranging from minor causes (flu, otitis, throat ache with fever, abdominal pain and diarrhea, with minor traumas as the most observed condition) up to life-threatening events. Besides, when it comes to children, parents know that hospitals can offer a variety of solutions (and possibly high-technological solutions) that are out of the reach of a GP surgery. On the other side, GPs and family pediatricians are supposed to know the child better, giving consequently better solutions.

In any case, acute conditions in children can be easily managed and permanently solved if a prompt and appropriate intervention is rapidly taken. Therefore, both approaches may have sense, provided that they are correctly accessed. How can the best of those be exploited, yet? As long as "telemedicine technologies involve real-time, live, interactive video and audio communication and allow pediatric critical care physicians to have a virtual presence at the bedside of any critically ill child [5]," the use of ICT tools can help doctors, parents, and children in acute and emergency settings.

The main applications of telemedicine for this purpose can be summarized as follows:

- Increase the diagnostic and treatment possibilities for family doctors (permanent connection that expands the capacity of a community surgery, with teleconsultation and online specialist referral, point-of-care device for hematological, biochemical, and instrumental diagnosis). Parents could perceive a level of care in a local clinic or surgery similar to the one given in the hospital. It is due to reduce the number of EDs' intervention [6].
- Apps that could help parents to self-assess the level of gravity of the child and to perform basic life support maneuvers if those are needed. It can be achieved with the use of:
  - Automatic advices
  - Specifically designed multimedia contents
  - Sensors already installed in common smart-devices as mobile phones or tablets that could detect basic life signs performing basic diagnostics or sending those data in remote receiving real-time feedbacks
  - Real-time consultations (doctors and trained health workers receiving data, audio and images, giving specific and tailored advices)
- ICT tools that could expand the level of care even for small centers, so that complex emergency can be managed also in small clinics with support and suggestions coming from bigger centers [7].

3.1.1 Family-Based Assistance

Families generally are aware of the risks of delayed interventions, both in high- and low-income settings. Yet many parents may not know how to manage an emergency situation, or because of the anxiety that such conditions commonly cause, they do not know where to go and who to contact. Most countries already have emergency numbers people can easily refer to. On the other end, in very rural areas, or in those places in which people do not pay sufficient attention to a child's health, the intervention can be delayed even with catastrophic outcomes.

Besides, some situations can occur when parents or caregivers are not at home. An ideal emergency service should be designed so that even children – in case of extreme need – can access it by themselves.

Telemedicine devices can support children and families offering a range of easy solutions that could help them to optimize the time and the efforts, enhancing the efficacy of the interventions.

Two main areas can be exploited:

- Offer guidance and tutorials that could help parents and children to perform basic life support maneuvers or other emergency maneuvers. It can be done thanks to *specifically* designed applications[1] or with the help of distant operators, specifically trained to give help in real time.
- Automatically or semiautomatically call for help and/or direct people to the nearest and most appropriate emergency center.

Both those systems can receive medical data from the remote connection, in order to rapidly assess the state of health of the child (tele-triage) and to establish which intervention is urgently required (distant support alone, video connection with a doctor, ambulance, ambulance with a doctor on call for an immediate home assistance, crash code call for the receiving hospital).

3.1.2 Emergency Teleconsultation at the Hospital

Another major problem is that not all the hospitals and the emergency units are able to provide the level of care needed for the management of acute and severe conditions in children. Most EDs, for instance, do not have a pediatric consultant 24/7 available or an anesthetist with a pediatric or neonatology experience. Some hospitals do not have a pediatric ward, and not every health center can give adequate support for special emergency situations (i.e., extensive burns, meningitis that requires isolation in a negative-pressure room, major traumas). Other hospitals are able to assist critical patients, but do not have the requirements to assist those same severe conditions in children [2].

The same areas of application that can apply to families can be extended to emergency departments and pediatric wards. The level of consultation switches from basic aid to professional assistance (the communication is between doctor and doctor in this case). Telemedicine application thus can offer:

- Assessment of the child's conditions, also with the acquisition of critical parameters or other physiological and biological parameters (both basic and advanced). Recorded or real-time data can be sent to a remote center (a specialist hospital, an advanced health center, a highly specialized centralized service acting as "call center" for health professionals[2]) where automatized systems or real-time consultations can help to define the picture. It can produce a telediagnosis, with crucial suggestions about how to manage acutely the patient. At the same time,

[1] Webpages and Facebook pages already give a number of unproven information easily accessible. Some of those sources, yet, are not reliable, with some of the procedures suggested resulting untruthful, wrong, not fit to the purpose information, and armful or risky when not performed by professionals.

[2] This is the case of the centers for the management of acute poisonings that already give real-time advices also to major hospitals, with telephone and online consultations. This service offers an interesting model: a very limited number of very specialist and specific centers serving all the national health system, from the smallest health unit to the most advanced one, also offering a call line for private users and families.

it can help to alert specialist centers where the specific condition can be managed and possibly treated.

- Guidance for those health professionals that are managing the acute scenario at a professional level. It applies to emergency departments and hospitals [8], but also to health workers and paramedics that assist the child in the scene of the event (for instance, an ambulance that reaches a scene of an accident where a child is involved and is in critical condition and therefore needs immediate help). Distant assistance can be given so that basic and complex procedures can be performed, in accordance with the minimum standard required for that peculiar condition and the capacity of the assisting team that is executing the intervention.
- The creation of a net among different health centers, so that the child could be directed to the most appropriate one, able to manage its specific medical condition at its best (it already happens today, for instance, when poisoning is suspected in a child: specific call centers with trained staff or websites give real-time consultations both for families and for professionals [2]). A health center with the minimum standard required can be automatically or manually identified and immediately alerted, and all the intermediate action can be consequently planned (alert the ambulance that has to transport the child to the place of intervention, call the doctors able to manage the case, ask for the correct blood type if needed, open the emergency operation theater and alert the surgical team if emergency surgery is required, alert radiologists and technicians, or contact the laboratory if special investigation is needed – rare microbiology cultures, for instance, in a suspect of a biological attack).
- In addition, the definition of a suspected case can automatically send alerts to other centers, so that if similar conditions have been reported in other emergency department, an epidemiology map could be immediately traced. This is particularly important when an outburst of an epidemic diseases is suspected (plague, cholera, hemorrhagic fevers, or other minor or major infectious diseases; food contamination; poisoning) or in cases of minor or major emergencies (biological or chemical terrorism or contamination, earthquakes, floods, or other natural disasters; other terroristic attacks).

In conclusion, telemedicine coordination can help to optimize the resource and to control the flow of intervention. This becomes crucial when child health is considered: in outbreaks and emergencies, children are the weakest subjects and the ones that mostly require specific assistance that not all the trained health professionals are able to offer [7].

3.2 Hospital Admission

Many consultations finally result in a hospital admission. There are three main reasons that bring a child to a pediatric ward:

- The identification of chronic or subacute conditions that need investigation that can be done only at the hospital level. It means all those investigations, procedures, and consultations that can be optimized and performed during a short- or long-time admission and those that cannot be done in a primary care setting (i.e., lumbar puncture, cardiac catheterization, electroencephalogram).

- The management of known medical condition that requires a 24-h surveillance or hospitalization (also in day-hospital regime) for investigation and treatment (e.g., the infusion of intravenous drugs or the reduction of the intrathecal pressure in children that suffer from chronic intracranial hypertension).
- The admission secondary to an acute onset of a disease or a trauma. In this case, it can be due to a direct act of will of the child or its next of kin (ambulance called at home by the parents or by the child, admission through a spontaneous presentation of the patient to an emergency department); to a request of deeper investigation asked by a GP or a community health professional; and to the automatic intervention of first aid teams in the scene of a critical event, for instance, secondary to a car accident.

Once the child enters in the hospital, two main different approaches can be followed, which depend on both the health systems and the kind of health center the child is admitted in (with a variety of possible situations between these two extreme endpoints):

- The acute condition is basically investigated and solved, or the required procedure is performed, then the child is discharged. Collateral or sub-conditions are not investigated and the further management of the child (when such an attention exists) is delayed in a further appointment with a pediatric consultant. This approach helps to reduce the waste and the burden of the hospital admission, but increases the risk of misjudged diagnosis and of medical complications, which can paradoxically result in an increment of the medical expenditure and of the burden for the patients.
- The child stays in the hospital until a reliable diagnosis is made or the medical condition is reasonably solved for good. This approach increases the level of care at its best, but produces an incredible expenditure that is likely to reduce the level of care in some other not properly covered areas. Besides, the number of days spent by a child in the hospital is increased, while other important activities (as school) can be compromised.

Telemedicine can help to create a middle point able to exploit the upside of the matter, reducing or eliminating the unfavorable terms naturally present in those two different situations.

Because health centers for secondary care and hospitals can be considered advanced medical facilities, the use of specific devices, designed for very specialist need, can be considered appropriate at this level of care. Yet, whereas the device itself has to be engineered in the most appropriate way in order to be as most accurate as possible (and consequently designed for very specific purposes and adopting the required technologies), the existing nets for communication or data management and processing should have to be exploited (see also Part II).

According to this profile, hospitals and health centers can take advantage of telemedicine for the following situations:

- *To optimize the flow of patients*, so that each condition can be immediately addressed to the correct facility and directed to the most appropriate professional.

It means both inside and outside the hospital itself (e.g., redirect ambulances with patients that have conditions that cannot be treated in a given hospital to the right medical facility; assist the triage phase and direct patient to the correct ward or surgery). Different tools can be used as GPSs (establish where the ambulance is and to which hospital to possibly refer the patient); smartphones or tablets can assist patients or paramedics to reach the correct area; telemetry can help professionals to assess the state of care of the patients before arrival in the hospital, so that the correct health team can be alerted or the ambulance can be rapidly redirected to other health centers if the hospital cannot guarantee the required level of care.

- *To perform a medical-assisted triage*: Complex cases and scenarios could take advantage of the distant consultation of a specialist consultant. Even before the child enters the hospital, pediatricians can assist the nursing team, evaluating in real time the state of health of the child and giving advices how to manage the case. It can be useful both in the small health center (i.e., a specialist from long-range connection can give advanced assistance) and in the major hospital complex (the physical extension of some facilities requires a considerable amount of time, before the call and the arrival of the specialist team).
- *To assist the pediatricians during the investigation of complex cases in the ward.* Specific devices can send information to highly specialized centers, and specialists can offer distant consultations; procedures can be assisted by distant professionals. It includes tele-surgery and robots controlled in remote. Consultation among different specialists can take place both in the same hospital and in connection with remote health centers.
- *To perform 24-h monitoring in the hospital.* Telemetry is already in use in the ICU with a central station receiving data from all the beds of the unit. This approach can be implemented also for semi-intensive patients or for those that are performing a 24-h monitoring for the investigation of specific conditions (ECG Holter in admitted patients). Doctors can receive information on portable device as tablets or smartphones, retrieving at the same time the patient's notes, other previous investigations, and other critical information.
- *To perform 24-h investigation at home.* Twenty-four-hour investigations that can be safely performed at home can use monitoring devices that can send information in real time (i.e., ECG or blood pressure Holter, polysomnography, apnea monitoring). Children consequently do not have to stay in the hospital just for a continuous monitoring of a health parameter needed to make a diagnosis. Children can continue the regimen of hospital admission at home, while parents can continue their everyday activities (alert can be sent both to the pediatric ward and to the parents at work, on personal devices and tools like smartphones, tablets, emails, SMS). It reduces the burden for children and families and the medical expenditure, and at the same time, it makes some investigations that are performed in real-life conditions more reliable (environmental bias is known to affect the gathering of data in some settings, secondary to the anxiety and of the stress that those same analyses produce and to the unnatural or hostile environment – as a hospital's room is for a child – where the examination is taken [9]).

- *Post-discharge appointment and follow-ups can be conducted on telemedicine basis*. Several tools can be used for this aim as video calls, telemetry, email and reminder, and specifically designed apps (see Chap. 5).
- *Prevention of complications and of relapses, rehabilitation, and psychological assistance* (see Chaps. 5 and 6)

3.2.1 Telediagnosis

The main goal for the pediatrician assisting a child is to make a diagnosis. This is a very simple concept that modern medicine sometimes forgets: specific conditions can undergo given procedures, following strict guidelines. This approach impacts on the symptoms and solves acute situations, but does not give a label to the disease the child is suffering from.

As we have seen, telemedicine can help health professionals to enter easily in touch among themselves, creating a net of horizontal and vertical consultations aimed to enhance the odds of success.

As medicine is becoming even more a complicated matter, the multidisciplinary approach is the most desirable solution to disentangle very intricate scenarios. Yet consultations are currently an issue, as long as specialists belonging to different areas are often unable to meet each other in order to discuss complex cases. Telephones in the past, and emails in more recent times, have been a valid instrument for professionals. ICT tools can offer today a valid support to enhance this experience.

The two main issues those devices are required to solve are as follows:

- *Help professionals to get in touch*: Apps and videoconferences can support the distant communication among doctors, optimizing the time and reducing the distances. Modern technologies allow to send composite information together with the voice or a video so that the experience can be improved going even beyond the face-to-face consultation (e.g., a digital radiography or CT scan can be discussed, during a video call, with the two consultants watching and operating in real time on the images or even on the imaging acquisition process). It also permits to use a multitasking approach.
- *Send and receive patient's data*: This is where research and development is mostly required. Medical data can be sent in telemetry to distant and highly specialized health centers, so that a real-time monitoring can be performed. It also allows delayed consultations by professionals that cannot give their specialist help at once. Besides, advanced investigations can be performed even in small centers. Specialized technicians can work in remote to complex devices, and data acquisition can be made even by personnel with minimum trained standards.

A teleconsultation might suggest a possible diagnosis, or specialists can suggest further specific investigations that fit a certain condition more. It is due to reduce the waste of resources needed for unnecessary examinations or the discomfort of some invasive procedures (e.g., a child with edema of the face, presenting in a small health unit and tele-referred to a pediatric nephrologist that based on the data received excludes the need of renal biopsy, suggesting less invasive investigations).

Those are not minor issues in pediatrics. Children are the ones that suffer the most, when parents roam from hospital to hospital trying to find the best assistance among different specialists. It delays the diagnosis and produces a significant quote of stress in children.

Besides, the use of distant diagnostic centers, in which specially trained professionals work only on the investigations or on the monitoring of a patient, without the need of a physical ward they have to work in, can help in the near future to reduce the costs of health, optimizing the resources and reducing the wastes. It implies more means for everyone and hopefully a better care. On the other side, this is a very risky approach, as long as it brings to the depersonalization of the medicine as well as the application of guided procedures that expose to medical errors.

3.2.2 Teletreatment

Similar issues lie beneath the need of treatments that can be done also in remote from distant specialists. Tele-surgery is the most remarkable application, allowing surgeons with top-level knowledge of given procedures to guide robots in remote also from intercontinental distances [10]. On the other hand, those are still very expensive applications that are not suitable for small health centers or low-income areas.

Medical treatment, administered under the supervision or according to the suggestion of a distant specialist, is another possible option, with less trained doctors, working in small health units, treating patients with therapeutic regimens that they are not acquainted with.

Yet there are other fields of interventions in which ICT tools can be used to deliver the treatment, both in the hospital and at home, after the discharge.

Children, indeed, have to learn how to take part in the management of their own health. In basic situation, when slighter illnesses are involved, this can be considered a minor issue. But when a chronic disease is diagnosed or a major intervention is required, the role that the child has is paramount. Diabetic children, for instance, or those kids that require continuous transfusions secondary to hematological problems often refuse the therapy or simply do not accept their condition. It is part of the management of the chronic child over time, but this rejection often starts in the hospital, once the diagnosis is made. ICT tools can help the children, since the early stage of the disease, to understand what is wrong with them and how they can be the main actors of their healing process.

Multimedia tools or psychological consultations can help the child to understand what is going on. Yet a main role can be given to Web 2.0 solutions in which many children and families with the same condition can enter in touch with other people with their same problem, learning from each other how to live with the disease (see also Chap. 6).

Reminders and rewards (e.g., trophy or small games that are unlocked every time a therapy is correctly taken or a progress in the rehabilitation process is achieved) are also an open way that telemedicine can explore (see also Chap. 5).

3.3 Telemedicine for a Family-Centered Care?

Although the recommendation is to implement a "family-centered care [11] for children," the countries in which a family pediatrician already works in the community are few. Thus, most of the specialist consultations still take place in the hospital. This represents a strong lack of opportunities as long as many appointments, needed to investigate an acute condition that possibly hides a chronic and still undiagnosed disease, are fixed after a considerable amount of time. Parents usually refer to the hospital for emergencies and acute diseases. In Western countries, the need for a rapid turnover needed to optimize resources and increase the hospital productivity is producing a reduction of the length of the admission. This can produce wrong diagnosis and medical errors.

This brief overview underlines the fact that even if most of the care for children should take place in the community and be hopefully home based and family centered, the hospital is still the place where parents are keen to go every time the child is sick. Yet hospital doctors have a very short opportunity to meet the child and have a close encounter with it. The whole process of gathering information is restricted to the medical encounter that in many cases lasts only a few minutes during the main round. Only complex conditions (or apparently complex conditions) are deeply investigated. Many children are only clerked by junior doctors and reviewed by seniors only at the time of the admission [3]. This is the time when most information are generally collected, but as long as many children get in touch with the hospital doctor for the first time during the admission (especially in major health centers), the accuracy of the anamnesis is low: no matter how deep the level of interrogation made when the history has been taken is, no hospital doctor can gather the number of information that a family doctor can potentially achieve over the years. General practitioners, in fact, should follow a child since its early years up to its adult life. It means that the whole medical history of a child can be collected through the years by the same doctor that is supposed to assess the patient continuously over time. This gives a full overview of the growing progress of the child, so that every time that a morbid condition shows itself or grows worse, an early detection can be made possible.

In some countries, as in Italy, children have a family pediatrician, sponsored by the national health system [12], where families can go for free. Its role is to assess the state of health of the child; to survey its growth up to the teen years, detecting and treating all those conditions that do not require for a hospitalization; and to investigate chronic and acute diseases whenever they appear.

Although this approach offers a valid support to the families, the risk of an excessive intervention and of an intensive medicalization of the childhood is constantly present.

The debate is still open, but a middle point among these two different approaches – the one mostly based in the community assistance and the one where the pediatrician works mainly at the hospital level – is to be sought.

But how can a hospital-based care transform itself into a more family-centered one? And how can a community pediatrician exploit the resources that only a major health center – as a hospital generally is – can offer?

Can the introduction of the telemedicine system help to achieve such a goal?

This is a very complex issue, indeed, as long as not only the function of telehealth has to be considered, rather the whole role of eHealth, starting from the implementation of the patient records or, in a wider meaning, the electronic health records (EHRs).

Yet several are the applications of telemedicine that can help to overcome the problem related to the management of a child's health at the hospital level:

- *Family doctor, hospital doctor consultation* [6]: Data required to formulate a diagnosis have possibly already been gathered in previous consultations, for those children with possible complex conditions that need deeper investigation. This happens because every encounter with a family doctor or with another specialist produces medical information. Not every chronic disease shows itself with acute onsets or with the instant blast that has produced the hospital admission. Most of them give minor discomforts that have been already investigated or treated in the past (celiac disease is a good example of such kind of situation). In addition, every swift from the normal growth process – if a child is continuously followed by community health workers, by an educator, or by the same family – can be already present in the health records based on previous consultations or on normal health assessments (for instance, a first visit asked before practicing a sport or a physical activity, a routine annual visit of a school doctor). Nowadays, hospital pediatricians are already used (or should be used to) to contact family doctors to gather further information whenever it is needed. Phone calls are a main resource for telemedicine, but they are based on an old-fashioned technology that does not exploit ICT tools at their best. Video calls, for instance, based on a direct and instant file and data sharing can enhance the level of consultation, so that the information shared can be completed and sound.
- *Access to electronic records by a hospital doctor*: On the other side, not all the information related to a child's health is present in a family doctor's database. Some families are not used to consulting their GP very often, for instance, and rather prefer to receive a direct consultation from an emergency department whenever the child has a problem. Some other families can change the place where they live a number of times over the growing years of the child, and for that reason, there are no family doctors that can produce a whole picture of the child's state of health or of its medical history. Because of that, medical information can be fragmented, even in a small child. Nevertheless, every consultation produces data, no matter where or when it took place. In an electronic-based health system, the implementation of EHRs is supposed to produce electronic directories where all the medical information of a patient can be stored and universally accessed. Those data can be read and processed by a distant user as

a hospital pediatrician during a specialist consultation, so that a complete picture of the child's state of health and medical history can be reconstructed. More information and more completed data can be crucial to achieve a diagnosis, especially in those cases in which the medical problem is blurred and the clues that could help to produce a conclusion are scattered over time and over the places.

- *Specialist center to specialist center consultation*: This is the main core of telemedicine, although not the only one (see also Sect. 3.2). Telemedicine can give to health professionals the possibility to enter in touch with the best specialists worldwide, receiving precious help and information that could help to investigate and manage also the most complex conditions. This is supposed to create a symbiotic process, in which the knowledge has no boundaries, where medical information and skills could be universally shared.
- *Family to hospital telediagnosis*: Once a chronic or subacute condition is suspected, whether that is secondary to a first hospitalization due to an acute onset or not, many investigations have to take place during the normal activities of the everyday life of the child. This is crucial to better evaluate the state of health of the child and to monitor the actual parameters, needed to confirm or reject a diagnosis. Some medical data (e.g., blood sugar level, heart rate, blood pressure) have to be accurately monitored to have important indications. When a chronic condition is suspected, some of those data have to be taken several times a day and possibly for more than 1 day. Today hospitalization is the only possible way to receive those information. Holter devices can also perform a similar chore, but still a hospital admission before and after the installation of those device is required, and the reading of those data is generally delayed. Yet when life parameters have to be controlled (e.g., an apnea monitoring in small infants), a real-time monitoring is required. The use of telemedicine devices, whether they are installed in a hospital or not, can help the pediatrician to make a diagnosis, allowing the child to live its life at home, reducing the burden, receiving more reliable information (the same hospitalization can produce a level of stress in the child that could undermine the accuracy of the tests [9]), and decreasing the time spent by the child in the doctor's office. The latter can also help to reduce the number of overall consultations in hospital and outpatient departments. This is likely to reduce the time that families have to wait to get an appointment with a pediatric consultant, increasing also the compliance to the follow-up.
- *Reminders (Telefollow-up)*: If a child requires for further consultations or investigations that do not need of a hospitalization and that can take place after the discharge, an appointment for a consultant referral is generally due. Yet many families do not consider it important to proceed with further consultations once the acute condition is solved, even if a chronic or complex disease is suspected underneath. Many are the possible reasons. As we have seen, the lapse of time that passes from the discharge to the consultant appointment can be one of these. The use of reminders can help the family to understand the importance of a follow-up and to organize their schedule, so that the appointment should not be missed (see also Chap. 5).

Conclusions

Hospitals are the natural place where care can be delivered to children and where advanced investigations and treatment can be done. Yet many are still the boundaries that limit the access to those services, and on the other side, many are the problems that come with a hospital admission, starting from those discomforts that undermine the quality of life of the child up to the waste of resource for unnecessary admissions.

The use of telemedicine devices, aimed to reduce the distances among the community and the hospital, to exploit the potentialities of small health centers and the in-depth knowledge of the child that family doctors have, to enhance the communication among different specialists, and to support families and children at home, has to be fostered.

The advanced equipment or the expertise of the specialists working for major health centers can be exploited also from those that live far from those locations, while children can receive an optimal care also within the environment they are more comfortable with.

Telemedicine implemented in the hospital system – which should not be disconnected to the community – can help in fact professionals to optimize their efforts, increasing the possibility of making an early diagnosis and selecting the best possible treatment; families to better face suffering, increasing the quality of care they can give to their children; and children to receive and seek better attention, reducing the burden.

References

1. Downing A, Rudge G (2006) A study of childhood attendance at emergency departments in the West Midlands region. Emerg Med J 23(5):391–393
2. Schappert SM, Bhuiya F (2012) Availability of pediatric services and equipment in emergency departments: United States, 2006. U.S. Department of Health and Human Services. Natl Health Stat Report. 1;(47):1–21.
3. Armona K, Stephensona T, Gabriela V, MacFaulb R, Ecclestona P, Wernekec U, Smithd S (2001) Determining the common medical presenting problems to an accident and emergency department. Arch Dis Child 84:390–392
4. Stephenson TJ (2000) Implications of the Crown report and nurse prescribing. Arch Dis Child 83:199–202
5. Marcin JP (2013) Telemedicine in the pediatric intensive care unit. Pediatr Clin North Am 60(3):581–592
6. Labarbera JM, Ellenby MS, Bouressa P, Burrell J, Flori HR, Marcin JP (2013) The impact of telemedicine intensivist support and a pediatric hospitalist program on a community hospital. Telemed J E Health 19(10):760–766
7. McSwain SD, Marcin JP (2014) Telemedicine for the care of children in the hospital setting. Pediatr Ann 43(2):e44–e49
8. Dharmar M, Romano PS, Kuppermann N, Nesbitt TS, Cole SL, Andrada ER, Vance C, Harvey DJ, Marcin JP (2013) Impact of critical care telemedicine consultations on children in rural emergency departments. Crit Care Med 41(10):2388–2395
9. Lewejohann L, Reinhard C, Schrewe A, Brandewiede J, Haemisch A, Görtz N, Schachner M, Sachser N (2006) Environmental bias? Effects of housing conditions, laboratory environment and experimenter on behavioral tests. Genes Brain Behav 5(1):64–72

10. Marescaux J, Leroy J, Rubino F, Smith M, Vix M, Simone M, Mutter D (2002) Transcontinental robot-assisted remote telesurgery: feasibility and potential applications. Ann Surg 235(4):487–492
11. American Academy of Pediatrics, Committee on Hospital Care, Institute for Family-Centered Care – Policy Statement (2003) Family-centered care and the pediatrician's role. Pediatrics 112(3):691–696
12. D. Lgs. 30 dicembre 1992, n. 502. Riordino della disciplina in materia sanitaria, a norma dell'articolo 1 della legge 23 ottobre 1992, n. 421. GU n.305 del 30-12-1992 – Suppl. Ordinario n. 137

4 Management at Home: The Chronic Child

Fabio Capello and Giuseppe Pili

Children who suffer from chronic conditions have disadvantages that go beyond the illness itself. Compared to other children of the same age, their quality of life can be highly compromised. It has been widely acknowledged that adult patients suffering from lifelong conditions can have difficulties in managing their own lives. The pool of opportunities or choices these people can take results compromised, as major life-changing decisions are often secondary to the health state [1]. The load to carry is so heavy that sometimes people experience a negation of their same condition or a rebellion against them that brings about the refusal of the same therapy. Besides, chronic conditions require chronic commitment. Those two reasons often lead to poor therapy compliance that may result – as in a vicious circle – in the deterioration of the health state and consequently the worsening of the burden.

But whereas adult people could have the strength, the experience, and the motivations to lead a battle against their disease, that could not be true for children. Besides, a child's life is not easy. Children with chronic disease are more subjected to harassment and to be bullied at school or can feel among themselves the heaviness of "not being like the other kids." Whole-time impairments are the gates of enduring stigma, able to compromise the child's physical, social, emotional, and psychological growing. Social skills, educational processes, behavioral attitudes, relationships, ordinary activities, cultural spurs, or future projections are therefore affected. The same life planning is put at risk, with the child unable to project himself or herself in the future, in the attempt to understand which choices he or she has left and which kind of man or woman he or she wants to become.

The burden of chronic diseases in children is summarized in Fig. 4.1.

F. Capello, MD, MSc (✉)
Pediatrics and Child Malnutrition, CUAMM – Doctors with Africa,
Via S. Francesco, Padova, Italy
e-mail: info@fabiocapello.net

G. Pili
Department of Child and Adolescent Psychiatry and Neurological Disorders,
ASL 1 Imperiese, Consultorio di Sanremo, Sanremo (IM), Italy
e-mail: giuspili@gmail.com

F. Capello et al. (eds.), *Telemedicine for Children's Health*, TELe-Health,
DOI 10.1007/978-3-319-06489-5_4,

Fig. 4.1 Quality of life in children with chronic conditions. Some of the most important causes of distress [3–5]

On the other hand, chronic illnesses largely vary, ranging from:

- Those conditions that require an in-depth monitoring in the early stage of a child’s life, but are likely to reduce over the years even when a strict follow-up for some time is needed (e.g., preterm infants)
- Those conditions that require continuous monitoring (e.g., blood sugar in children with type I diabetes)
- Those conditions that need specific therapies that can be delivered at home (e.g., management of acute attacks or prophylaxis in children affected by severe asthma) or exclusively at the hospital (e.g., blood transfusion in children affected by thalassemia major or dialysis in children with chronic renal failure)
- up to those conditions that give specific impairments or major physical limitations (e.g., cardiac disease that limits the physical activity of a child; chronic conditions that expose the child to an augmented risk of disease, as, for instance, *osteogenesis imperfecta*; physical handicaps as paraplegia that limits everyday activities; major genetic syndromes that completely disrupt the normal way of life)

In addition, there are situations related to the state of health of a child that could be limited over time and space, but that can have strong consequences in its future life. There could be in fact:

- Diseases that bring in themselves the same issues related to lifelong conditions, but that can impact on a child’s life mainly during a specific range of time, but that can possibly resolve themselves after a long treatment

- Conditions in which children would grow out of them
- Conditions that can leave minor sequelae even after the clinical healing
- Conditions that required a lifelong or a time-limited follow-up (cancer being an example that can summarize all the previous options)

Interestingly, some of the burdens related to a chronic condition do not lie in the disease itself. Some studies show how patients and families consider the therapy as the most important factor for the reduction of the quality of life of the child [2].

Children with disabilities, besides, are often apt to become the target of depreciable behaviors from kids of their same age. This is a major concern that can deeply affect the life of a child and compromise its future psychological and social development.

Another major issue is the lack of information that strikes the families at the moment of diagnosis. Parents and children are particularly sensible that they can gen up starting from any possible material available online or off-line. Often the research aims to find alternative cures, able to revive the hopes. Yet this process often brings to the gathering of wrong information from unreliable sources that could lead parents to wrong decisions (as suspending a treatment) or to dangerous behaviors (as experimenting an unlicensed drug, scraped from the Internet) or even worse into the web of frauds and charlatans [6].

To make things worse, chronic conditions can come as well in association: a number of children experience a complex chronic condition (CCC) that distorts their lives and – in major impairments – also the ones of their next of kin. Those families spend a considerable amount of time in hospitals and in consultations or on the way to reach them. Those are deep life-changing experiences that affect all the course of a family's life. Home-based treatments and follow-ups could help to increase the quality of life, acting on a major cause of distress for those people. Unfortunately, the patients that exploit home assistance are still few [7].

Assessing the quality of life for a child with chronic disease thus is not easy [8]. But here is where telemedicine can contribute to reduce the burden, helping children and families to exploit the care and conduct a good life.

4.1 Telemedicine for the Screening and Early Diagnosis of Chronic Diseases

Discovering the existence of a lifelong or life-affecting disease is a sad experience that starts from the diagnosis. Those diseases that are congenital can be discovered at a very early stage (secondary prevention) in order to augment the success of treatment or rehabilitation in those cases where an early intervention can be crucial or to prepare families to the challenges they will find ahead. Genetic screenings and prenatal screenings [9, 10] are a common practice nowadays. The standard adoption of software that can be used in the doctor's office, at secondary care level, or delivered directly at home through the Internet can contribute to spread the reach of those technologies. In fact, clinical decision support (CDS) software is likely to change the way those screenings will be conducted.

As long as many chronic diseases are not congenital and some are merely related to environmental factors, early interventions and diagnosis are crucial. This is particularly true in those critical situations when proper investigation and diagnosis can deeply affect the outcomes (e.g., triage and multidisciplinary approach in major traumas in children). Telemedicine can extend the resources normally available in the points of care: point-of-care devices and supporting decision tools (real-time consultation from remote) are possible options. The aim is the building of a net that can augment the odds for a doctor – virtually to the excellence – in assisting a complex patient. That is due to improve the level of care and therefore of the outcome [11]. Assessing correctly the patient since the onset of a disease or just after a trauma took place can in fact prevent the development of future chronic condition. The early intervention and the prevention of complications – which are secondary to a good diagnosis – can prevent as well the development of conditions that can have a major impact in the future life of a child. Those indeed are not only related to major events but also to those situations that can be referred to as minor or non-life-threatening disorders or traumas (e.g., facial laceration that needs a prompt and proper surgery [12]), but that can have lifelong sequelae.

Moreover, as chronic condition in children can be due to environmental process (especially in genetic susceptible children), telemedicine has to have a key role in prevention. As we have seen, ICT tools can play a lead part in primary care (see also Chap. 3). Prevention and prophylaxis can be promoted with mobile and Web-based alerts and health educational programs that can start from the school and can continue at home, reaching a considerable number of children that today are even nearer – more than their counterpart, the adults – to the "always-connected" reality (see Chap. 12).

4.2 Telemedicine for the Treatment at Home

The two major areas of intervention for a home-based care are related to the reduction of the burden and the improvement of the compliance. That can be achieved with the analysis of the major problems that contribute to affect the quality of life in children and in their families and in those that undermine the adherence to the treatment. Home-based intervention – when integrated with a specific support – has been proved to be effective and sustainable [13].

Another important issue is related to the spread of hospital-acquired infection (HAI), to which chronic children are more prone (both because of increased exposure due to the recurrent admissions and to their precarious health conditions that made them more susceptible to infections than in healthy children). Reducing the number of hospitalizations, thanks to home-based treatments monitored in remote, can decrease the risk of secondary infections that can complicate the basal condition.

4.2.1 Better Treatment, Lesser Strain

As we have seen, children have to deal with a number of issues when they have to go through lifelong or long-lasting treatments. Reducing the burden of those could certainly help to improve their quality of life and to increase the adherence to the treatment that eventually will lead to an even better life. Telemedicine can offer a spread of solutions that could help children to follow a treatment without experiencing the bad of it.

Obesity in children is a field of application where telemedicine has been introduced with a good outcome [14] and can offer an example of the management of a chronic condition in which a proper intervention can restore the quality of life of a child. Telemedicine devices can play a major role as they:

- Can create a continuous stimulus to the child that can be supported by relatives, friends, and professionals in the management of its own condition (distant care)
- Can be alerted whenever there is a shift in the optimal response and adherence to the treatment (monitoring in real time and from home of the body weight, the food intake, the physical activities, the social life, and the reinforcements needed to improve the compliance). This can be achieved already today with the use of mobile technologies and smart-devices, able to automatically detect some of the parameters needed to evaluate the treatment (as the miles covered on foot every day thanks to GPS detectors built in on a mobile phone or the heart rate checked in continuous by a smartwatch) and to wirelessly transmit them to a central database (that can be accessed by the doctor for assessment and evaluation of the adherence and the response to the therapy, at home for parent monitoring, and by the child for continuous feedback aimed to consolidate positive behaviors)
- Can offer a net of support (Web 2.0 technologies) among children and families that suffer from a similar condition or to ill and healthy children in an attempt to reduce the distances among them
- Give prompt assistance in case of harassment and bullying

Besides, telemedicine can help doctors to understand why a specific treatment is not accepted by a child, understanding and implementing those behavioral and cognitive interventions that are required to improve the compliance [15].

4.2.2 Telemedicine for Rehabilitation

Many chronic conditions are not related to the disease in itself, but are due to a long course that starts when the health condition arises (a diagnosis or an acute onset that gives complications and sequelae) and that could end when a rehabilitation process is properly conducted (tertiary prevention). A child that suffers from the multiple lesions of a road accident, for instance, can experience a very long healing process that can eventually end with a total recovery, but that could need a constant and continuous commitment. The risk of failure along the way – especially for children that could taste the diversity in their lives in a more sensitive way than adults do – is extreme: children may simply give up because they do not want to undergo a boring and discriminant process, the goal of which is way out of their sight.

Children have to go through a complex and unpleasant process that could last years and that could increase the risk of failure or relapses just because the child at some point decides to give up. Reducing the burden means also improving their quality of life during the process.

On the other side, children can be easily involved in funny activities that they could see more as a game rather than a chore. Video gaming, for instance, has been proved useful for physical rehabilitation even in adults [16, 17]. Using a tool that children already know and that are apt to use (also because of its enjoyable contents) can enhance the adherence. It can result even more useful if those same activities can be shared among children worldwide that are experiencing the same problem.

Therapy and rehabilitation probably will never be a leisure for children, but they can at least assume the face of it.

4.3 Telemedicine for Monitoring and Follow-Up

A key factor in the management of the health of a child with a chronic disease is the real-time monitoring of the child conditions and of its response to therapy. Besides, a continuous feedback between the pediatrician and family helps the child to achieve a better compliance. This is not a secondary issue as long as many children affected by lifelong diseases fail to continue their treatment once they become teenagers, even when they are perfectly aware that their same life is in jeopardy [18].

The WHO already states why interventions are needed in order to improve the adherence suggesting some crucial issues [19] that telemedicine could be able to address:

> Poor adherence to treatment of chronic diseases is a worldwide problem of striking magnitude. The impact of poor adherence grows as the burden of chronic disease grows worldwide. Poor adherence to long-term therapies severely compromises the effectiveness of treatment making this a critical issue in population health both from the perspective of quality of life and of health economics. Improving adherence also enhances patients' safety Because most of the care needed for chronic conditions is based on patient self-management (usually requiring complex multi-therapies), use of medical technology for monitoring, and changes in the patient's lifestyle, patients face several potentially life-threatening risks if not appropriately supported by the health system. Adherence is an important modifier of health system effectiveness Patient-tailored interventions are required There is no single intervention strategy, or package of strategies that has been shown to be effective across all patients, conditions and settings. Consequently, interventions that target adherence must be tailored to the particular illness-related demands experienced by the patient. To accomplish this, health systems and providers need to develop means of accurately assessing not only adherence, but also those factors that influence it. Adherence is a dynamic process that needs to be followed up. Improving adherence requires a continuous and dynamic process. Recent research in the behavioural sciences has revealed that the patient population can be segmented according to level-of-readiness to follow health recommendations. The lack of a match between patient readiness and the practitioner's attempts at intervention means

that treatments are frequently prescribed to patients who are not ready to follow them. Health care providers should be able to assess the patient's readiness to adhere, provide advice on how to do it, and follow up the patient's progress at every contact. For the effective provision of care for chronic conditions, it is necessary that the patient, the family and the community who support him or her all play an active role. Social support has been consistently reported as an important factor affecting health outcomes and behaviours. There is substantial evidence that peer support among patients can improve adherence to therapy while reducing the amount of time devoted by the health professionals to the care of chronic conditions. A multidisciplinary approach towards adherence is needed: A stronger commitment to a multidisciplinary approach is needed to make progress in this area. This will require coordinated action from health professionals, researchers, health planners and policy-makers.

Diabetes can be considered a paradigm in the use of mobile and ICT devices for the monitoring and treatment of chronic children. Telemedicine devices have already been proved effective in the following applications for the management of this heavy and impairing condition as "the automatic transfer of blood glucose data; health education based on the use of SMS; management of a diabetes diary for diabetic patients and a food picture diary; integrate a patient diabetes diary with health care providers; monitoring of physical activity; give information on nutrition e special diets; context sensitivity in mobile self-help tools; and modeling of blood sugar using mobile phones [20]."

Some conditions require a first stage of continuous and strict surveillance for a limited number of months or years: preterm infants, children with a very low weight at birth, for instance, normally need a 2-year follow-up after the discharge. That period of time can become longer in case of major sequelae (e.g., brain injuries, retinopathy, or bronchopulmonary dysplasia). The use of videoconference devices or apps designed for the Web can be a possible solution to reduce the need of medical consultation in the hospital [21], lightening the burden for families and children.

A possible solution to enhance the quality of home-based monitoring and follow-up could lie in compact portable devices for analyses. Wireless multiparametric analyzers should test one or more values using the similar technologies now available for home-based monitoring (like glucometers for glycemia, INR detectors, or blood test analyzer for the hematic value of cholesterol or triglycerides). Those are simple and relatively cheap devices (also known as point-of-care [POC] devices or portable blood analyzers) that could easily test one or more values and that could integrate an automatic system able to connect the device to the Internet and send the retrieved data in real time to a patient personal record in remote or to a doctor for an online evaluation of the state of health of the child [22, 23].

Two possible POC devices can be described:

- Personal devices that parents can buy or lease from the hospital and that stay with the child (proper home-based monitoring and follow-up); those are compact and simple-to-use devices that require a minimal technical knowledge (one-touch devices), possibly no need of maintenance and use of disposable materials (as the strips with reagent needed to collect the blood sample and test the needed value).

- Expensive and more sophisticated POC community devices that can be used for a number of children in neighbor-based health centers. This could be particularly useful in those rural areas (see Part III) in which health centers and hospitals are out of reach and children (especially those that need a continuous monitoring of therapy and of their medical conditions) cannot access healthcare on a regular basis.

4.4 Telemedicine to Inform and Support Children with Chronic Diseases and Their Families

Most of the burden that lies beneath chronic conditions is related to the psychological sufferance that affects children and families and goes further than the expenses related to drugs, transportations to and from the hospital, days of works or school lost, and so on. Those costs are difficult to evaluate, but are probably the ones that have the most important impact on a child's and on its family's life (see also Chap. 6).

Information and support can be exchanged and shared among specialist and families. Besides, telecommunication systems could help families to share useful information among themselves [24].

Moreover, the chronic child is bound to become a chronic adult. Many patients in addition can lose their way in the transition from childhood to adulthood. Without a proper intervention and coordination and a specific support for the families, the quality of care can be deeply affected. This has been also highlighted in some medical conditions as the hematological diseases that require a lifelong treatment and follow-up: "although comprehensive programs exist in paediatrics, affected adults may not have access to preventative and comprehensive healthcare because of a lack of providers or care coordination. They are often forced to rely on urgent care, leading to increased healthcare utilization costs and inappropriate treatment [25]."

ICT tools have a crucial role in both providing prompt assistance in case of need (e.g., teleconference that bypass the need for scheduled appointment or can safeguard the child's privacy; real-time messaging service that could offer advices and support in case of need; open and anonymous connection with organization offering aid, legal, and psychological support in case of child harassment secondary to its medical condition) and helping to create a net of contacts among children and families that share the same problems (Web 2.0-based tools with restricted access for members in order to guarantee the privacy of children and families, linked with a net of professionals that could offer mediation in real time).

The wide spread of Internet-based devices can help children to easily access those tools, creating a preferential way of communication among the different actors involved. Children who nowadays are more familiar than the adults with ICT tools can use a channel that they already know and master, in order to help themselves to get help.

Conclusions

Telehealth can have a primary role in the management of the chronic child, both monitoring and supporting the treatment and improving the adherence to therapy, helping therefore families to receive a continuous help and support. It is due to reduce the burden for children and families, whose quality of life is often strongly affected by the medical condition affecting the child. The idea of a home-based treatment that can reduce the time of hospitalization and the visits to a medical surgery does not sacrifice the doctor-patient relationship, which can be kept reducing the distances that can normally occur in a conventional medical encounter. A net of professionals and of families sharing the same problems – based on easily accessible technologies powered by the Internet – can improve the compliance, helping children to overcome those issues that normally undermine their quality of life.

References

1. Bhatti ZU, Finlay AY, Bolton CE, George L, Halcox JP, Jones SM, Ketchell RI, Moore RH, Salek MS (2013) Chronic disease influences over 40 major life-changing decisions (MLCDs): a qualitative study in dermatology and general medicine. J Eur Acad Dermatol Venereol. 2013 Oct 18. doi:10.1111/jdv.12289. [Epub ahead of print]
2. Knowles RL, Day T, Wade A, Bull C, Wren C, Dezateux C, UK Collaborative Study of Congenital Heart Defects (UKCSCHD), Adwani S, Bu'lock F, Craig B, Daubeney P, Derrick G, Elliott M, Franklin R, Gibbs J, Knight B, Lim J, Magee A, Martin R, Miller P, Qureshi S, Rosenthal E, Salmon A, Sullivan I, Thakker P, Thomson J, Wilson D, Wong A (2014) Patient-reported quality of life outcomes for children with serious congenital heart defects. Arch Dis Child 99(5):413–419
3. Nakou S (2001) Measurement of quality of life in the health care field. Applications in child birth. Arch Hellen Med 18(3):254–266
4. Daliento L, Mapelli D, Volpe B (2006) Measurement of cognitive outcome and quality of life in congenital heart disease. Heart 92(4):569–574
5. Nousi D, Christou A (2010) Factors affecting the quality of life in children with congenital heart disease. Health Sci J 4(2):94–100
6. Rinaldi G, Gaddi AV, Capello F (2013) Medical data, information economy and federative networks: the concepts underlying the comprehensive electronic clinical record framework. Nova Science Publishers Inc, Hauppauge. ISBN 978-1-62257-845-0
7. Lindley LC, Lyon ME (2013) A profile of children with complex chronic conditions at end of life among medicaid beneficiaries: implications for health care reform. J Palliat Med 16(11):1388–1393
8. Upton P, Lawford J, Eiser C (2008) Parent-child agreement across child health-related quality of life instruments: a review of the literature. Qual Life Res 17(6):895–913
9. Hilgart JS, Hayward JA, Coles B, Iredale R (2012) Telegenetics: a systematic review of telemedicine in genetics services. Genet Med 14(9):765–776
10. Edelman EA, Lin BK, Doksum T, Drohan B, Edelson V, Dolan SM, Hughes K, O'Leary J, Vasquez L, Copeland S, Galvin SL, Degroat N, Pardanani S, Gregory Feero W, Adams C, Jones R, Scott J (2013) Evaluation of a novel electronic genetic screening and clinical decision support tool in prenatal clinical settings. Matern Child Health J. doi:10.1007/s10995-013-1358-y
11. Dharmar M, Romano PS, Kuppermann N, Nesbitt TS, Cole SL, Andrada ER, Vance C, Harvey DJ, Marcin JP (2013) Impact of critical care telemedicine consultations on children in rural emergency departments. Crit Care Med 41(10):2388–2395

12. Farook SA, Davis AK, Sadiq Z, Dua R, Newman L (2013) A retrospective study of the influence of telemedicine in the management of pediatric facial lacerations. Pediatr Emerg Care 29(8):912–915
13. Carter B, Coad J, Bray L, Goodenough T, Moore A, Anderson C, Clinchant A, Widdas D (2012) Home-based care for special healthcare needs: community children's nursing services. Nurs Res 61(4):260–268
14. Davis AM, Sampilo M, Gallagher KS, Landrum Y, Malone B (2013) Treating rural pediatric obesity through telemedicine: outcomes from a small randomized controlled trial. J Pediatr Psychol 38(9):932–943
15. Hommel KA, Hente E, Herzer M, Ingerski LM, Denson LA (2013) Telehealth behavioral treatment for medication nonadherence: a pilot and feasibility study. Eur J Gastroenterol Hepatol 25(4):469–473
16. Szturm T, Betker AL, Moussavi Z, Desai A, Goodman V (2011) Effects of an interactive computer game exercise regimen on balance impairment in frail community-dwelling older adults: a randomized controlled trial. Phys Ther 91(10):1449–1462
17. Laver KE, George S, Thomas S, Deutsch JE, Crotty M (2011) Virtual reality for stroke rehabilitation. Cochrane Database Syst Rev. (9):CD008349
18. World Health Organization (2003) Adherence to long-term therapies: evidence for action. ISBN 9241545992 (NLM classification: W 85)
19. Windebank KP, Spinetta JJ (2008) Do as I say or die: compliance in adolescents with cancer. Pediatr. Pediatr Blood Cancer 50:1099–1100. doi:10.1002/pbc.21460
20. Årsand E, Frøisland DH, Skrøvseth SO, Chomutare T, Tatara N, Hartvigsen G, Tufano JT (2012) Mobile health applications to assist patients with diabetes: lessons learned and design implications. J Diabetes Sci Technol 6(5):1197–1206
21. Gund A, Sjöqvist BA, Wigert H, Hentz E, Lindecrantz K, Bry K (2013) A randomized controlled study about the use of eHealth in the home health care of premature infants. BMC Med Inform Decis Mak 13:22
22. Kost GJ, Curtis CM (2012) Optimizing global resiliency in public health, emergency response, and disaster medicine. Point Care 11(2):94–95
23. Lee BS, Lee YU, Kim HS, Kim TH, Park J, Lee JG, Kim J, Kim H, Lee WG, Cho YK (2011) Fully integrated lab-on-a-disc for simultaneous analysis of biochemistry and immunoassay from whole blood. Lab Chip 11(1):70–78
24. Oprescu F, Campo S, Lowe J, Andsager J, Morcuende JA (2013) Online information exchanges for parents of children with a rare health condition: key findings from an online support community. J Med Internet Res 15(1):e16
25. Kanter J, Kruse-Jarres R (2013) Management of sickle cell disease from childhood through adulthood. Blood Rev. doi:10.1016/j.blre.2013.09.001, pii: S0268-960X(13)00057-X

5 Overtaking the Distances: The Child with Special Needs

Giuseppe Pili

Compared with the children of their same age, kids with disabilities are commonly considered disadvantaged. Physical, mental, social, or psychological impairments are conditions that affect deeply their quality of life, undermining at the same time the chances of their adult life. The price that those children have to pay is augmented by the distrust or the pity they arouse in those people that surround them and their families.

Living with a disability means carrying on your own shoulder the distance that average people cast on yourself. This is an added burden that goes beyond the disease in itself, but that is able to affect the quality of life of a child, more than the condition by which it is affected of.

Besides, children have to interact with boys and girls of their same age that are often not used to deal with diversity unless a fertile and positive education has been offered to them.

On the other hand, children can be considered different not only in relation with a specific health condition but also because of differences related with their social status, ethnicity, religious belief, and the economic and political background of their families and their emotional responsiveness and the capacity to build and maintain relationships. Moreover, children live complex lives, with own rules, in which the onset of unseemly behaviors is part of their normal approach to the world.

In addition, the same Internet-based technologies suffer of three main phenomena that can challenge a child's life:

- The Internet addiction
- The cyberbullying and Internet-based harassment
- The handiness of unsuitable contents for children

G. Pili
Department of Child and Adolescent Psychiatry and Neurological Disorders,
ASL 1 Imperiese, Consultorio di Sanremo, Sanremo (IM), Italy
e-mail: giuspili@gmail.com

F. Capello et al. (eds.), *Telemedicine for Children's Health*, TELe-Health,
DOI 10.1007/978-3-319-06489-5_5, © Springer International Publishing Switzerland 2014

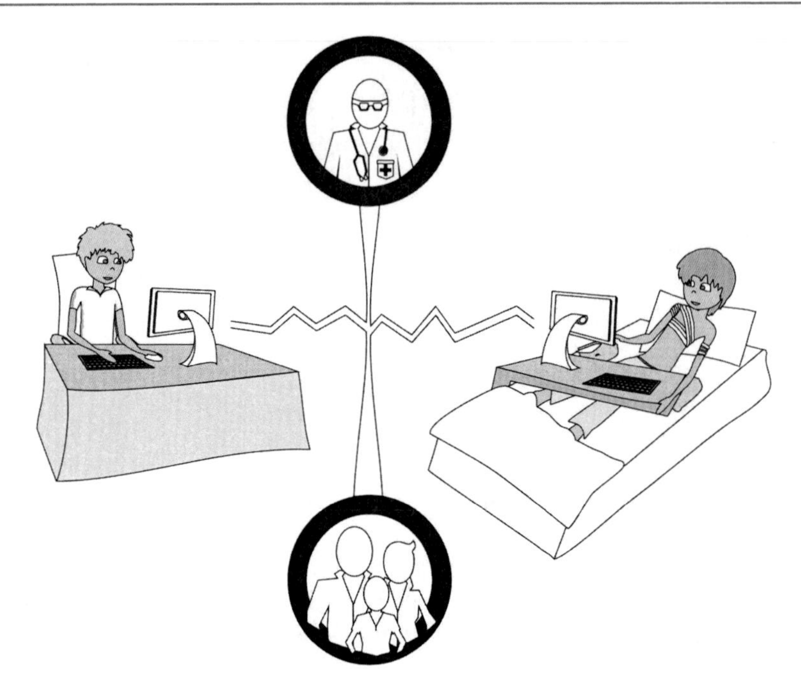

Fig. 5.1 Kids affected by disabling condition should be encouraged to get in touch with other children of their age, in order to reduce the burden of their disease. A net of people, where health professionals and families can be part of that communication, could ease the process and detect in real time discomfort, anxiety, and those possible threats that can come from the Internet or from an irresponsible use of connecting devices

It implies that those same technologies discussed so far – able to offer a way to reduce the distances among children with impairments and healthy children – can lead to disastrous results when their use and implementation is unwary.

Nonetheless, the use of ICT tools and the prompt action of all the actors involved in the care, starting from the families, can prevent the establishment of those problems and help in the early detection and management of such inconvenient behaviors.

The aim is the construction of a net of people, children and adults, that concur together to the wellness of the youngsters, in order to fill the gap between healthy and impaired children (Fig. 5.1).

5.1 Telehealth to Detect Awkwardness and Discomforts

Some possible application of communication technologies and consequently of telehealth can be used to early detect discomfort in those children that are affected by chronic or impairing conditions or in those that are particularly vulnerable and in need but do not ask – for different reasons – for help.

5.1.1 Detect Discomfort in Children (With or Without Disabilities)

Children browse the Internet every day. The number of underage people that make use of connecting devices or that check social networks on regular basis is constantly increasing.

Children submit queries on the search engines, post user-generated contents and comments, and receive and send messages to people from around the world.

The cathartic appeal of social network and Web of the 2.0 generation helps children to express themselves also delivering to the Internet all those feelings and inner emotions that they are generally unable to show in their private and everyday life.

Software running on the Internet has already been proved to offer a possible aid in the mapping of peculiar health-related events [1].

The use of those tools can be translated in those apps or browsers children use to communicate, in order to detect in real time any kind of discomfort that children are unable to communicate to parents, friends, teachers, or health workers.

In a communication model in which the child should exploit ICT devices and the Internet to achieve a better health (and that is particularly true for children suffering from chronic conditions, disabilities, or impairment), the development of tools that could help to detect this discomfort as soon as possible is crucial. There could not be a telemedicine for children unless the inner life of the child is safeguarded and its emotional experience is protected.

Nevertheless, the privacy of the child has to be protected. It does not only mean that personal data cannot be freely shared and published without the consent of the family or of the child but rather that children have the right to have their own private world, in which grown-ups should better stay away from.

5.1.2 Prevent Harassment and Bullying Among Children

The implementation of a net of services that could help the child to get in touch with other kids of their age or to stay connected to the world even when its disability limits its actions has to deal with a number of threats that come from the Internet itself. One of the major issues that can affect the serendipity of a child using connecting devices lies in the increment of episodes of harassment (also known as cyberbullying) among children making use of social networks.

This is not a minor issue, as long as an increasing number of reports are demonstrating how depression, anxiety, and suicidal thoughts can be detected in children that are victims of these assaults [2–5].

The most vulnerable ones are those same special children that are likely to be shut out in the real world. It means not only children with physical, social, or psychological impairment but also those very sensitive kids that could be easily targeted.

Therefore, there could not be a theory of telehealth for children if those current issues are not being taken into account. A monitoring system, mediated by parental guidance and the help of professionals, is paramount. In particular, the role of professionals, properly trained to deal with children and with those specific issues, is crucial, as long as the little victims of bullying are usually not keen to discuss their problem with their friends or families. It is a silent suffering the one these children are affected of. A double strategy should be based on a push information model (easily accessible contents sent to the children when discomfort has been strongly suspected) and on automatic interventions (alerts that detect possible problems and offer to the children – in the same browser or app they are using – links and connections to professionals able to give tailored help and support in real time; automatic alerts sent to health professionals when an abuse or a harassment is strongly suspected). It also implies that children especially when they made use of technologies intended to support their health (telemedicine software and device) have to be encouraged (or the model has to be planned in order to get this result) to browse in close nets or apps, where these safety tools can run. Other antivirus-like apps or software can also be studied in order to detect in real time criminal behavior that could affect the health of those children that made use of the Internet (for health, for study, or for leisure).

5.2 Telehealth to Offer Support

Some of the steps that Web 2.0 technologies can help to achieve vary from the simple request of information from people that already suffer from a particular disease up to a real-time consultation with help and advice givers. A net of users, sharing the same issues and linked to a body of professionals that could mediate those relationships, can help families to find the best strategies and mostly to avoid the devastating sensation of being alone, which is probably what makes chronic disabilities often unbearable.

Besides, children have their own way of communicating that is peculiar. When it comes to support or health education (which is the first step to prevention and that is due also and especially for those children that are already affected or susceptible to be affected by chronic or debilitating medical conditions), this has to be considered an opportunity: online video games [6] and role-play games, for instance, can help children to build their net, so that knowledge can be spread, support can be offered and sought, and alerts can be bleeped as soon as some discomfort is arising.

Some of the common advices to help parents to better understand how to manage a child with disabilities (Table 5.1) can easily find a solution in fact with computer-based and telemedicine devices. Information given and shared, besides, can be stored in local or in the Cloud, so that alerts and reminders can be given whenever it is needed. Those can be delivered regardless of the physical location and to different users (to a child, or to its teacher, at school and at the same time to a parent at work), in a dynamic process in which all the services provided by telehealth are linked together in order to work jointly and not as separate entities unable to communicate among each other.

Table 5.1 Some of the advices that parents generally receive when they have to deal with a child with a chronic condition or a disability [7]

When you look for a specialist with whom you can work, ask other parents of children with disabilities. Often, they can give a good recommendation or speak about their experience. That could help to find the right solution or to deal with the management of a child disability more confidently
Sometimes doctors do not realize that what is common for them cannot be common for people that do not come from a medical background. Always ask for clarification, especially if you do not understand the words used by doctors or health workers. Do not be afraid to say "I don't understand." You are not a doctor and it is your right to ask for clarifications
When needed – especially when some therapy has to be given to a child – write down the answer given by doctors and health professionals
Do not deny your child's condition. Try on the contrary to find as much information as you can about your child's condition. It can help in communicating with other people that are experiencing the same problem, help to understand what doctors and health professionals are doing or are trying to explain, and help you to feel part of the team that is taking care of your child
Prepare a list of the questions you would like to ask or discuss to doctors or therapists before the visit
Keep cautiously the notes of your child's health: the medical history, the hospital's admissions and the discharge letters, the test results, and any other information that could help you or the doctor to manage your child's health. A diary-like note can also be useful to evaluate changes and progressions
Do not be afraid to talk with professionals and to disagree with them if you feel like it. Just remember to clearly explain your discomforts and your motivations
Do not give up your life. The child would suffer more. Take care of your own schedule, finances, and other commitments and try to live as more normal as you can. Medical recommendations and treatments have to fit in the normal life of the child and of the family and should not have to be the linchpin of your existences

Some of them are really useful because they come from people that have experienced the same problems and had already gone through the same pains. The possible risk is the delivery of unreliable information or of information that applied to a child but that does not fit with the medical condition of another one

When it comes to rare and very rare diseases, it is crucial as long as the sensation of being alone or being the unlucky one among millions is devastating.

Tables 5.2 and 5.3 show the major cadres of disabilities and the most frequent conditions that create impairment in children.

5.3 Telehealth to Educate People and Help Children with Disabilities to Reduce the Distances

The use of Internet-based technologies has already created a net of people that spread worldwide across the continents, regardless of the physical location or the diversity in terms of culture, social status, income, or languages. Pictures, video, and multimedia known as user-generated contents broadcasted on the main networks as YouTube are shared every day through social networks like Facebook, Instagram,

Table 5.2 Categories of diverse disabilities [8]

Developmental disabilities
Intellectual or cognitive disabilities
Physical disabilities
Hearing impairments
Visual impairments
Deaf-blindness
Traumatic brain injury
Speech and language disabilities
Learning disabilities
Mental health disabilities
Emotional disturbance
Orthopedic impairment
Multiple disabilities

Table 5.3 Some common conditions bound to create distances among children and families, undermining at the same time their quality of life

Genetic syndromes and neurological conditions	Toilet habits (soiling, bed-wetting and daytime wetting)
Epilepsy Muscular dystrophy Osteogenesis imperfecta Cerebral palsy	Autism ADHD Learning disabilities Stepparenting
Celiac disease	Arthritis and rheumatologic conditions
Diabetes	Osteoskeletal dimorphisms
Chronic anemia	Posttraumatic or postsurgery impairments
Asthma	Cystic fibrosis
HIV/AIDS	Transplanted children
Cancer	Speech disorders
Cardiac diseases and malformations	Deafness
Child abuse and neglect	Dumbness
Substance abuse and addiction	Metabolic syndromes
Sexually related concerns	Growth disorders
Developmental delays	Chronic pains and skin lesions (also secondary to medical procedures)
Mental diseases Depression Anxiety Bullying and child harassment Obesity Birth defects	Bowel malformations and chronic diseases

or Twitter. Besides, knowledge also from reliable sources is easily nowadays available from the Internet, thanks to wiki-like websites that offer information that can be peer-reviewed in real time or to professional pages (universities, government and nongovernment organizations, nonprofit associations, and scientific publications that offer guidance for parents, kids, and professionals available for free download). Such a wealthy resource is already contributing to spread information that can help parents and children to understand and manage their conditions and help families to get in touch with other people (both healthy and affected from the same condition) and professionals.

The goal to reach in the near future, nevertheless, is the certification of the contents available on the Internet and the construction of channels for people with children affected by specific disease, so that a net of reliable resource can be built. This is crucial, considering that the Internet is still considered as one of the most important sources of false, misleading, and unreliable information that could lead to very dangerous behaviors in terms of self-assessment and self-drug and therapy prescription [9, 10].

On the other hand, ICT is not only intended to share experiences and support sensitization campaigns and advocacy but also to provide advice and solutions.

If the same idea of health is not the mere absence of illness, the role of telehealth is to improve the quality of life of children and families, overtaking those physical, social, economic, and emotional limitations that are mainly the cause of distress in an ill child. In other words, it means that – thanks to the already available technologies and hopefully to the ones that have not been invented still – children can get in touch with other children, in spite of their own impairment or physical condition. As a big leveler, the Internet can provide senses and means that go beyond the physical restraints. Touching a screen or acting on a mouse can actually be a very small action for a kid, but it can certainly produce giant leaps for the children worldwide.

Conclusions

Children with disabilities or those ones that are most sensible and susceptible to other people's harassment suffer from a chronic pain that goes beyond the illness in itself. The idea of being alone is devastating especially for children. Kids with permanent impairments never had the possibility to experience a life comparable to the ones of the other children. They feel like the ones that had picked the right numbers of the wrong lottery.

The use of communication technologies able to reduce the distances among those children and families, together with the implementation of a governed net able to support those same children and detect the abuses, is a possible option.

The aim is to create a safe playground in which children could feel children again in spite of their condition or their physical location (at home or in the hospital, for instance), trying to recreate those conditions that the impairment they are suffering from has erased from their lives.

References

1. Ginsberg J, Mohebbi MH, Patel RS, Brammer L, Smolinski MS, Brilliant L (2009) Detecting influenza epidemics using search engine query data. Nature 457:1012–1014. doi:10.1038/nature07634
2. Daine K, Hawton K, Singaravelu V, Stewart A, Simkin S, Montgomery P (2013) The power of the web: a systematic review of studies of the influence of the internet on self-harm and suicide in young people. PLoS One 8(10):e77555

3. Bucchianeri MM, Eisenberg ME, Wall MM, Piran N, Neumark-Sztainer D (2013) Multiple types of harassment: associations with emotional well-being and unhealthy behaviors in adolescents. J Adolesc Health 54(6):724–9
4. Stewart RW, Drescher CF, Maack DJ, Ebesutani C, Young J (2014) The development and psychometric investigation of the cyberbullying scale. J Interpers Violence 29(12): 2218–2238
5. Kindrick K, Castro J, Messias E (2013) Sadness, suicide, and bullying in Arkansas: results from the Youth Risk Behavior Survey – 2011. J Ark Med Soc 110(5):90–91
6. Wilkinson N, Ang RP, Goh DH (2008) Online video game therapy for mental health concerns: a review. Int J Soc Psychiatry 54(4):370–382
7. Gaddi A, Capello F, Manca M (2014) eHealth, care and quality of life. Springer, Milan, 30 Dec 2013. ISBN 8847052521
8. Rinaldi G, Gaddi A, Capello F (2013) Medical data, information economy and federative networks: the concepts underlying the comprehensive electronic clinical record framework. Nova Science Publishers, Hauppauge. ISBN 1622578457
9. Brown C, Goodman S, Küpper L (1992) When you learn that your child has a disability. Originally published 1992, 2nd edn, 2003. Updated, 2010. NICHCY (National Dissemination Center for Children with Disabilities). Available from: http://nichcy.org/families-community/journey. Accessed 19 Jan 2014
10. Categories of diverse disabilities. ACTion sheet: PHP-c128 (2004) Available from: http://www.pacer.org/parent/php/PHP-c128.pdf. Accessed 19 Jan 2014

Suggested Reading

Barrett Singer AT (1999) Coping with your childs chronic illness. Robert D. Reed, San Francisco. ISBN 1885003145

Batshaw ML (2001) When your child has a disability: the complete sourcebook of daily and medical care. Paul H. Brookes Publishing, Baltimore. ISBN 1557664722

Bedwetting. ERIC's guide for parents (2012) - ERIC (Education and Resources for Improving Childhood Continence), Bristol Also available from: http://www.eric.org.uk/assets/18512%20ERIC%20Bedwet%20Guide%20red.pdf

Boyse K, Boujaoude L, Laundy J (2014) Children with chronic conditions. UMHS (University of Michigan Health System). Nov 2012. http://www.med.umich.edu/yourchild/topics/chronic.htm. Accessed 19 Jan 2014

Centers for Disease Control and Prevention (2003) Special focus: health-related quality of life. Chronic Dis Notes Rep 16(1):1–36

CG111 Nocturnal enuresis – the management of bedwetting in children and young people: NICE

clinical guideline 111 (2010) NICE (National Institute for Health and Clinical Excellence). NHS, London. ISBN 978-1-84936-369-3 Also available from: http://www.nice.org.uk/nicemedia/live/13246/51382/51382.pdf

Gillespie A (2008) Child exploitation and communication technologies. Russell House Publishing Limited, Lyme Regis. ISBN 1905541236

Helping kids deal with bullies. Reviewed by: D'Arcy Lyness. Kidshealth. Date reviewed: July 2013. http://kidshealth.org/parent/emotions/feelings/bullies.html. Accessed 19 Jan 2014

Huegel K, Verdick E (1998) Young people and chronic illness: true stories, help and hope. Free Spirit Publishing, Minneapolis. ISBN 1575420414

Sam's story (2010) Illustrations by Sally Flynn. ERIC (Education and Resources for Improving Childhood Continence)

U.S. Department of Health and Human Services (HHS) (2014) 2020 topics & objectives: maternal, infant, and child health. http://www.healthypeople.gov. Accessed 20 Jan 2014

Part II

Technical Issues

6 Connectivity, Devices, and Interfaces: Worldwide Interconnections

Andrea E. Naimoli

Telehealth and distant medicine mean above all communication, namely, creating a relationship among different actors (humans and/or machines) placed in different locations. One of the first problems to face is the interconnection of subjects, with two main roles to manage: a more active one, traditionally the doctor or the health worker, and a more passive one, usually the patient. Interconnection in itself is made up of two components: a transport medium and an interface.

Nowadays, worldwide communications are vastly available, being a common support in all developed (technologically) countries as Web connections and mobile phones. Besides, GSM and satellite connections are becoming widely available also in developing and least developed countries, with many portable solutions already under implementation [1]. Synergies among Wi-Fi systems and portable devices are under development [2] – for low- and high-technological areas, as long as the importance of accessible Wi-Fi connections also for those settings has been stressed since the beginning of the Web 2.0 era [3].

As a first-step analysis, we set a good medium and interface choice at a high level of abstraction: in the most common scenario – a typical situation – the transport medium is the Web [4], while the interface is a Web-connected device, such as a notebook or a mobile phone. We must state that in developing countries, these facilities are often unavailable at the moment or had a limited spread and ineffective maintenance (see also Chaps. 9 and 10). Nonetheless, the aim is to work and build around this scheme, as it would be the cheapest and fastest way to cover all needs [5].

This is a standard, affordable, and sustainable approach, but it can be deceptive (lack of 24/7 power supply, poor-quality connection, malfunctions, breakage of vital components with no available replacements or spares, poor coverage also due to hostile environmental conditions, lack of trained technician for setup and

A.E. Naimoli
Tech Department, Airpim Inc., Wilmington, DE, USA
e-mail: andrea.naimoli@elementica.com

F. Capello et al. (eds.), *Telemedicine for Children's Health*, TELe-Health,
DOI 10.1007/978-3-319-06489-5_6,

maintenance, wrong handling of fragile components, misuse of the systems also for inappropriate scopes[1]).

What if we cannot achieve this result then? As for the transport medium, we will refer to the nearest reachable node (the nearest Web-connected terminal) with an additional effort to carry prominent data with less conventional systems (it means a pen drive in the best case or even a paper sheet to be scanned afterward!), while for the interface, we are assuming that at least a simple terminal is available, as long as telemedicine models necessarily require a minimum threshold of technologies for the implementation of basic functions. Nonetheless, considering the worst possible scenario, traditional systems again can be taken into account, in the same way we do for its transport counterpart. In the undesirable situation where connection is a challenge (no presence, low, or very low quality), additional transports have to be considered just a kind of "tail" in the communication model. In our studies, we will refer to it as an element of noise, typically that which slows communications down.

After all, the state of the art in communication worldwide is reaching a very high standard, with a set of Web-connected nodes all around the world already available and accessible, either with direct-linked devices (in the best case) or with a kind of indirect-linked points (in the worst), the speed of transmission being the only sensible difference among them. It should be stressed that for direct-linked points, we can manage real-time communications, while for indirect-linked ones, we must take time gaps into account. For direct-linked points, we can have a more or less stable connection too: at the end of the day, what we have to consider is that a "node" should be considered always linked (with no difference between direct and indirect ones), but characterized by a τ factor proportional to the speed transmission. In a specific context, we can also define an arbitrary or computed ideal top-speed value T and consequently an efficiency attribute of a node, namely, η, as the division between τ and T.

τ = node's speed transmission (average)

T = context (arbitrary ideal or computed average) top reachable speed

η = node's efficiency = τ/T

In real-time communications, T could be in the order of a few hundreds to a few thousands of milliseconds even for great distances (totally at the odds too). Just to focus, a realistic situation could have T = 3,000 ms (at most 1 s of delay) [6].

It is important to understand that speed is often part of a wider context: if a medical test requires about 2 days to be performed, there would be no point in receiving the results back in 10 seconds or in 10 minutes in 10 s or 10 min (+6,000 %), being such a lapse of time critical, on the other hand, is if the test is quite instant (heart rate and life signs monitoring in a critical scenario, as in a mobile emergency department – major emergencies or humanitarian settings – or in a tele-monitored intensive care unit).

[1] e.g. the waste of data traffic for improper use as unauthorized downloading of multimedia contents from the net, so that the weekly or monthly ceiling (commonly applied for disadvantaged connection, as the ones in rural and extreme rural areas, especially in developing settings) is reached before the time. Professional use of the connection system becomes thus compromised.

Connectivity is the infrastructure, the workbook. Devices and interfaces (i.e., hardware + software) are the tangible players in the game, the pencil to write in the workbook: these must be felt as friendly and usable by people.

6.1 Targeting the Subject

Design and development of devices and interfaces must consider their target: to whom are those instruments designed for? medical and specialist doctors? patients? illiterate people? experts? The first dichotomy we want to emphasize in this work is "adults versus children."

Many elements diverge in targeting a child or an adult, such as:

- Level of instruction and general knowledge (including the knowledge of signs and symbols that may not still be acquired in children as a green color for a good result and a red color for a wrong one)
- Practical experience (mainly to react to unexpected events or react proportionally to them, e.g., inconsolable burst of cries when a painless medical maneuver is performed by the doctor)
- Physical measures (e.g., size of fingers: this could be a point for touch devices)
- Physical strength (e.g., to carry a complex tool to monitor some parameters)
- Motivation
- Visual acuity
- Emotional involvement (including also those irrational behaviors that are typical of the younger age groups, e.g., fear of a stethoscope or of an ultrasound machine, both painless, noninvasive, and harmless)
- Magical thought (lack of cause-effect connections in the child's thoughts)

Just to be clear, children will be more comfortable with well-known shaped devices like a smartphone, instead of a complex strange instrument seen as a "medical tools," or with a friendly shaped equipment (see also Fig. 6.1).

Some of the main features for a device for telemedicine include:

6.1.1 UIs (User Interfaces)

UIs are the subsequent challenge, not reinventing the wheel again, but focusing on established technologies. A wide range of alternatives is on the way, having to choose between:

- *Output*: textual/graphical representation
- *Input*: pseudo-free/guided interaction
- *Content*: descriptive/conceptual contents

Again, the optimal choice has to be a workable and fit-for-use interface that children could understand and/or not fear.

Fig. 6.1 Dealing with objects children are familiar with can help to reduce the stress of the medical encounter. Properly designed interfaces, based mainly on charming illustrations and graphics could provide a more engaging experience. The aim is to provide a valid tool aimed to reduce the stress of the medical encounter, exploiting all the opportunities that come from the telemedicine approach. Besides, captivating apps and devices are likely to incept healthy behavior in children in a more effective way while giving feedback with a different level of complexity (in the example, according to the value recorded: a laughing or a sad face for children, a green or a red light for parents, the specific value sent in remote to a health center)

6.1.2 Output Choice

In childhood, the use of pictures is prominent [7] and a graphical output seems to be the way to go: after all, GUIs – graphical user interfaces – are the often first choice for home and personal computing.

6.1.3 Input Choice

Input processes (we mean "how to reach and activate an option": multilevel menus, an open shell to enter commands, and so on) are very important. More features bring more complexity in the system. If we want total freedom, a full command line system[2] would be an optimal solution: this is a coder-side approach, not suitable for common people. On the other hand, we can use graphical representation for the available actions, but as long as they provide less freedom, they must be chosen wisely.

[2] A *command-line interface* (CLI) is a coding console where the user types commands to achieve desired results: this is a kind of programming, and it could be very hostile. Its counterpart is the *graphical user interface*, where almost everything is a graphical object, the most common being a WIMP (windows, icons, mouse, menus, and pointers) environment.

6.1.4 Content Choice

Speaking of contents, we have to underline the desired level of detail we aim to achieve: for a possible diagnosis, we can print out a full list of technical names if the user is a doctor, but for the patient, a shortest description in terms of "practical effects" should be better. Optimal systems should be able to adjust the complexity of the contents showed according to the audience (namely, the final user), without losing the nature of the message (progressive simplification of a subject, starting from fit-to-fact content, e g., point-to-point plot of the variation of the blood sugar level in the monitoring of type I diabetes, to a figurative and therefore stylized representation of the same – a calendar with happy face or sad face days according to the value of blood sugar recorded over the week; see also Fig. 6.1).

6.2 Device Ergonomics

Digital information cannot be stored or transmitted without the support of a physical device. Therefore, we should consider what kind of hardware we need or we can count on. We can start considering "static" workplaces, as is the case of hospitals, health centers, or GP surgeries (where the care is traditionally delivered): not-moving workstations could be set up using a common personal computer. But as long as telemedicine is intended also for remote, not so comfortable places or, moreover, for patients who need continuous monitoring or interaction or those who have mobility limitations, alternative solutions have to be explored. Here is where the so-called portable category comes to action. Some clarifications, though, have to be done: first of all, it should be obvious that an object can be considered portable for a man but not for another (as in the case of a child that is naturally weaker and smaller than a grown-up); then a very important subset has to be emphasized, namely, related to meaning of a *wearable* one.

6.2.1 Portable Devices

We define *portable* as any device that the user can carry around, therefore not so bulky or heavy to impede its translation, both with or without the use of secondary tools needed for its transportation (first, mobile phones could be considered portable, as long as they do not need to be wired to a fixed connection, but they were so heavy that generally they were built in into a car console).

6.2.2 Wearable Devices

This is a subset of the previous category; consequently all the features considered for *portable* devices still apply: their peculiarity is that they can also be "linked" someway to the user himself or herself, usually as a kind of wear. We take a list of some examples starting from possible adaptable wear objects that can be used as wearable devices and then with a list of some components that can be integrated in the stated wears.

- Wear objects:
  - Bangle
  - Necklace
  - Wristwatch
  - Belt
  - Shoes
  - Glasses
- Components:
  - ICC (integrated circuit card)
  - RFID (radiofrequency identification) chips
  - Passive tags (as QR codes)

Wearable devices are less prone to be forgotten and more comfortable to be adopted: they are perfect for continuous monitoring and can become a seamless habit for everyone.

6.2.3 The Energy Problem

Electronic devices need energy to work. This is a basic statement that has to be cautiously considered as long as most of the telemedicine services in order to be reliable need to be potentially 24/7 operative or reachable. The choice is between cabled and wireless devices, the latter improving the mobility and the wearability, but needing an outside source of power or of a periodical (re)charge. In any case, a reliable source of energy has to be present and always available, both for wired or wireless devices. There is a range of possible technical solutions that can be adopted; the most optimal ones have to be chosen by developers taking into account the characteristics of the settings the devices are supposed to work in.

6.3 Architecture: Local and Remote

Considering the whole context of worldwide, Web-supported interconnections, we identify the local actors, such as the doctor and his young patient, and the remote ones, such as Web-based servers where those services rely on (this is the case of a very common and simple Web search, made with a search engine like Google). In our effort of creating a seamless, efficient integration of all systems, the way data is stored, transmitted, and received becomes critical. Having the Web as the platform, a two-way transmission can use very well-established standards. Nonetheless, this does not apply for data storage, where many formats are available.

There are also other technological aspects with a variety of alternatives that should have to be addressed, but format (in the sense of "file format") is a crucial one, as long as it can be considered as a sort of "language" that all the involved actors have to understand. On the contrary, other elements, like interfaces, programming languages, and operating systems, can interoperate if wisely used: as a person can handwrite a text in his or her own writing style, provided that he or she uses a comprehensible language that the reader knows and understands. According to this example, the file format is the language, while the writing style is all the rest.

6.3.1 Data Format

Rather than a technical choice, we need a philosophical one: the choice of a good reference model on how storing data is a start point that needs to evolve time by time. Without reinventing the wheel, the Open Data Model is a possible way to go, accompanied obviously by open format specifications: adoption of public documented formats is in substance the matter that counts. Not to be too vague, a simple example is the XML technology, but any other with similar power would be good: XML allows to virtually represent any kind of information (i.e., text, graphics, multimedia) with any level of complexity. Besides, the structure of interrelated pieces can be defined. Basing our communications on the present Web, we could imagine that the future will be something like the Web 3.0.

Conclusions

New technologies can be an affordable way to improve everyday life and patients' care, provided that they are used wisely, starting from the assessment of the real needs and of the restraints that lie in the nature of the final users (in this case the child). A layer of complexity can be created in order to offer different solutions for different users, without compromising the nature of the communication or of the contents.

Any device intended for children has to consider the peculiarities appropriate for this group of users and also consider the wide variability among the different kids and age groups.

The technical limitations of the setting those devices are intended for have to be carefully evaluated in order to avoid a waste of resource and the introduction of impracticable models that are due to delay the proper implementation of telehealth systems.

References

1. Customer Case Study. Vodafone and Linksys 3G/wireless router opens new market and demonstrates strategic collaboration. Cisco Systems, Inc., IBSG (2006) Available from: https://www.cisco.com/web/about/ac79/docs/wp/Vodafone_Littlebox_CS_v4.pdf. Accessed 10 Mar 2014
2. Ghadialy Z. The growing synergy between small cells and Wi-Fi. Available from: http://smallcells.3g4g.co.uk/2013/08/the-growing-synergy-between-small-cells.html. Accessed 10 Mar 2014
3. The wireless internet opportunity for developing countries. Edited by The Wireless Internet Institute, also in behalf of infoDev Program of The World Bank, United Nations ICT Task Force. Geneva, 2003 and following updates. ISBN 0-9747607-0-6

4. Berners-Lee T, Fischetti M (2000) Weaving the Web: the original design and ultimate destiny of the World Wide Web by its Inventor. HarperBusiness, New York. ISBN-13: 978–0062515872
5. Adam Hyde (2008) The Contributors of FLOSS Manuals OLPC laptop users guide. The Contributors. Berlin. p 153. ISBN 9780615260679
6. Mills DL (1983) Internet delay experiments. IETF, Fremont, California. Available from http://tools.ietf.org/html/rfc889. Accessed 10 Mar 2014
7. Schocken Book (1996) Lillard PP Montessori today: a comprehensive approach to education from birth to adulthood. ISBN 978-0805210613, New York

7 Technology and Social Web: Social Worldwide Interactions

Andrea E. Naimoli

Considering the general scheme we are focusing on, we should note that when we connect a device to another Web-linked one, that same device becomes wrapped as well, as part of an oneness: the huge Web, connecting today more than three billion people worldwide. This is an opportunity to be taken. We can define for every given person (whether he or she is linked or not to the Web) a Web profile:

Web profile	Profile of a person as a set of available or computable information in the Web

With that, we do not mean a profile that is generated once someone registers to and therefore creates an account on a social network (SN): in fact it is not necessary to use the Web, to be inside the Web. People live in the Web whether they want (or like) it or not. To understand this concept, we can just think of all the historical figures who lived before the Web era. They never had the opportunity to create their own Web profile, as long as the Internet or computer technology when they were alive was not even invented. Nevertheless, we have tons of information about them. Besides, today those information are also supplied with metadata that univocally related the historical figure with the proper one and not with a homonym (a Wikipedia entry about Leonardo da Vinci, for instance, collects narrative and quantitative data that univocally refer to the man that painted the Gioconda, plus metadata – connected with sublinks that univocally refer to other voices, as the clickable picture of paintings made by him – that give a univocal and specific meaning to the entry itself [1]).

As in the "real" world we are someone, the same applies to the "virtual" one. Moreover, our virtual alter ego becomes part of ourselves, so that in the Web, we expose only a fraction of our complete essence (see Fig. 7.1).

A.E. Naimoli
Tech Department, Airpim Inc., Wilmington, DE, USA
e-mail: andrea.naimoli@elementica.com

F. Capello et al. (eds.), *Telemedicine for Children's Health*, TELe-Health,
DOI 10.1007/978-3-319-06489-5_7, © Springer International Publishing Switzerland 2014

Fig. 7.1 The information that corresponds to a single person exists in a physical world (i.e., actions, documents, photos, letters, ID cards, school or medical reports, awards, newspaper cuts, interactions with other people) that can be defined as "real" and, in electronic datasets, generated by people's everyday interaction with electronic tools and the Web [2] that contribute to create the "virtual" image of our alter ego

In technical words, this profile is composed of complex data: we need to summarize how these data are made available, so we set the following definition:

Web profile's element	Digitally representable information, with a computable value having custom granularity and visibility

We will see in Sect. 7.2 how these elements work, but it is crucial first to state now the importance of the concept of granularity [3] and what granularity actually is: *we want each element to have a value, but with a tunable precision and a known delta of error in precision.*

All those features and consequently the information stored in single profiles can be made available to different users and for different scopes. This would be possible when we consider that different layers of the information can be shared, each one of them adding details and going in depth to the core or to the whole complexity of the information itself. It means that different services can be delivered or different users could retail data according to the aim they access those data for. The same information could be blurred or accurate (see Fig. 7.2) when, for instance, confidentiality is needed (an epidemiological survey may want to know how many children missed school days in the previous school year, but it would not be interested in knowing who and why they did not attend school. A school doctor instead has the authorization to access those data, without breaking the confidentiality, in order to understand why a specific child is skipping school and how often).

Web-based services can consequently be implemented, open the access to data – in a governed way – to third-party providers of such services.

In the following paragraphs, we see how we can exploit social interactions, presenting as well some with some innovative proposals.

Fig. 7.2 The owner of the information could access the integrity of the data and can choose to release those details he wants to share with third parties. The ones who access the data, accordingly to the level of authorization – and secondary to specific purposes – can see it in a more blurred or in a more accurate way, with or without receiving details regarding the identity of the owner of the information itself

7.1 Breaking the Language Walls

Apart from strictly medical needs, ICT is very useful for collateral actions, such as language understanding: in a worldwide context and in situations where doctors and patients coming from different cultural backgrounds – and therefore different native languages – an incompatibility of communication is to be expected. This issue has to be addressed in a telemedicine system that virtually should connect every part of the world, so that connection is due to happen among very distant places (in geographical and in cultural terms). Social Web can offer big advantages in order to fill this gap.

Four technological features that can be combined to obtain the expected result are hereby listed: instant translation, speech recognition, writing recognition, and speech synthesis. A proposal based on those characteristics and related to those considerations would follow.

Instant translation

Nowadays, instant translation of written text is as simple as accessing a Web service or using a small smartphone: although not perfect, it can be considered fairly correct, with the major errors taking place most for less spoken languages.

Nonetheless, the general meaning of the conversation is generally quite understandable, with few or very few information lost in translation. The first step in managing interlanguage communications is the adoption of such a solution.

Speech recognition

Another feature in this field is the so-called speech recognition: spoken words become written (digital) text. Once again, this is an already available technology as the previous one. It could be very useful to store a natural speech as a digital medium without the overwork of transcription.

Writing recognition

The so-called writing recognition is the capability of an ICT system to convert natural writing in digital data: it is becoming more and more powerful, so that today it is available also for low-cost smartphones.

Speech synthesis

The latter point we speak about is *speech synthesis*: a digital text is translated in audible form as spoken by a human; being available for many years (even the 1980s home computer Commodore 64, 1 MB of RAM, 1 GHz of CPU, was capable of it!), it has been improved greatly over time, so that today it is sometimes difficult to distinguish synthesized from real voices, with the most advanced system already able to mimic the regional accents.

7.1.1 The Proposal: Real-Time Free Instant Translation

A feasible proposal, aimed to reduce the cultural gap among the two actors in a medical encounter or of a telehealth connection, comes next: through a dedicated Web service we are already working on, instant translation, not only for written but also for the spoken language, is available, funding its resources on a global community of volunteers.

A model case could be an English-speaking doctor treating an Xxx-speaking child: using a very affordable device, such as a Web-linked smartphone connected to the above service, they can communicate with few efforts, each one in his own language. This would be a cultural revolution, especially for children's health, as long as it would make possible the communication among children that usually are able to understand and speak only one language, and potentially every specialist around the world.

Besides, it would boost the level of communication among different health professionals despite of the cultural background they came from. Although many medical doctors speak or understand at least written or simple English, many health workers do not. This would be crucial as well in emergency scenarios, when the availability of a health professional is reduced and the work force could not be short-listed only to the ones that are able to communicate with foreigner parties.

Especially for medicine, and for pediatrics, in which some sets of standard questions and answers also supported by visual material can be created, this additional tool would increase the possibilities for doctors and caregivers, augmenting the potential applications and benefits of a telemedicine system.

7.1.2 Costs

A small note about costs: as any technological choice, all solutions reported in this book have a cost. This is not the aim of this work, but we just point out that this approach is likely to be much less expensive than any other similar solution for the majority of the situation. In addition, in a number of cases, it demonstrates itself much more effective. Of course, this is not always true and the quality varies a lot, accordingly also to the application and the final purpose, but as a temporary, quick solution, that statement is quite ever a must.

7.2 Interactions

The (maybe) most obvious kind of interaction is a peer-to-peer one even at long-distance level, with doctors directly getting and keeping in touch with patients (in case of specific diseases, for instance, a child could carry with him a mobile phone or a smart device, so that in case of emergency, he can call the doctor or his family). Interaction can as well happen indirectly (parents, relatives, nearby persons, teachers at the school, educators, schoolmates, coaches, etc., reporting an episode related to the child or asking for immediate intervention).

Avoiding long-distance interaction that necessitates the child to physically go to a specific place, dislocated far from his normal environment, is anyway one of the first goal to reach. This point is to be emphasized: for children and families, long traveling and time spent for traveling are a major burden [4], and it is a bigger problem for disadvantaged people, for those living in rural areas [5]. The same applies to children with physical impairments and special conditions, where telemedicine can play a major role in delivering easily accessible consultations [6].

In other words, when no physical contact is required, why walk for kilometers just to have a meeting when the doctor is just a click away?

7.2.1 Data Storage and Analysis

Adoption of technical devices leads to another opportunity: collection of data. From the point of view of a single, we have the chance to create a long-lasting medical record to be kept safe for present and future uses: this is especially true for children, where such records could be built starting from scratch (although some bureaucratic concerns could arise, limiting the use in those groups of age), in order to create a complete electronic record able to store the integrity of the child's health-related issues. The same consideration applies for some realities in developing countries, especially in those settings in which a proper national health system has not been properly established yet.

From the point of view of the doctor, there is another chance available: the collection of data can result in a source for analysis, both in response to the needs of the moment (and therefore for their immediate use) and in behalf of the global community (e.g., in evidence-based medicine).

Nonetheless, many are the restraints and the considerations related to the designing, the implementation, and the sharing of electronic health records, and such complexity has to be managed in the wider idea of eHealth [7].

7.3 Social Privacy

We cannot go further without speaking of privacy, as long as we have seen how easily data moves around, going forth and back, up and down, here and there: anyone has the right to choose which information to upload on the Web and in the system and above all in a shared one and how those information would be collected, inputted, managed, processed, shared, and mined for future and still unknown uses (provided that he or she has been informed on the consequences either positive or negative).

Privacy is a key issue especially for older children that many will share some information of their own life, but only to a specific counterpart, being shamed to reveal some of their problems to other parts (problems related to sexual development, for instance, in preteens that are not keen to share these questions with their relatives or specific chronic problems that the child does not want his mate to be acquainted about, as bed-wetting or soiling).

We have seen that a Web profile is composed of a set of elements where we can identify at least a value, a granularity, and a visibility: let us consider a very simple one, for instance, the body temperature. This element can be defined by a simple subset of contents, like instant (hour and day) and level (Celsius degree), that are the basic data.

Starting from this example – the storage of a simple (or a complex) data – there are a number of issues that have to be addressed:

What to store: the value

First of all, the subject – the child or (it depends on the age) his tutor (maybe a parent) – decides if he wants to store digitally a specific information. This can be easily achieved by manually inserting values in a custom software, even if a kind of automation is possible, linking a custom digital thermometer to the Web or with the child having a tablet with a fit-for-purpose app (e.g., a sort of game where touching an appealing image the software asks the child to enter this data). In the management of chronic conditions, the active involvement of the child is a key point to improve the adherence to the therapy and the compliance to the monitoring and follow-up. This simple gesture, consequently, has already achieved results.

Note that we have done nothing more than storing the data: we know not still how it will be retrieved.

What precision: the granularity and the visibility

If body temperature is 36.3 °C, we can choose a granularity of 0.5 °C on an integer scale: this way retrieving the value, it will appear as to be 36.5 °C (rounded to the nearest 0.5). The idea here is that we can define a set of granularities, based on the retrieving way, so as to have 0.1 °C for us, 0.25 °C for the hospital, and

0.5 °C as a public information. What is happening? When the owner of the information retrieves it, it will have the best precise response possible, but when it came to a doctor or a nurse that does not need the accurate data (but may be more interested in qualitative information; the child is feverish or not? Is the temperature high or very high?), that value could be rounded somehow. Moreover, the same information could be made public, for example, for statistical purposes. Note that in this case, only temperature is made public and not the owner's identity.

Consequently, when we store information, we should have the possibility to define how these can be retrieved: we associate for each "channel of retrieving" a granularity level.

Relations of data

Speaking of visibility, we underline that it is not just a matter of "yes/no" flags: all information is related to each other, so – taking the example of the body temperature back – we do not simply state "my doctor CAN'T see my temperature" or "the world CAN see my temperature," but something that goes from "my doctor CAN'T see my temperature" (no visibility at all) to "my doctor CAN see my temperature BUT with a granularity of 0.5 °C" (partial visibility) or "my doctor CAN see my temperature BUT without knowing who I am" (partial visibility again, but different from the previous one) to "my doctor CAN see my temperature knowing who I am" (full visibility for the doctor), and so on.

7.4 Social Community

The community is the set of all Web-profiled people: there could be active (they do interactions) or passive (e.g., the ones whose data are inserted by others and do not use any of the facilities available) users. In this network of subjects, there is a wide cloud of information with references and accesses, led by the rules set as the global social privacy (Sect. 7.1.2): data are somewhere, but not everyone can access them and not always with the same privileges. Usually the owner has the greatest access on his own data, but this is not always true: children are a typical example, as there should be a kind of tutor (hopefully a parent or a relative, but could be a friend or any other delegated – by law – subject) to manage all needs, at least up to the legal age.

How those communities are supposed to work and support the child's health, improving the chance for acute and chronic children to stay well, reducing the distances, and trying to neutralize the collateral effects (or early detect them) of the Web on children, has been summarized in Chap. 6.

7.4.1 Natural Living

In addition, we should recall in memory how things work in the past and still work today in some developing areas or rural villages: children went playing around, outdoor, with a load of advices but also a sort of independence. In case of emergency, who was the one involved? This can appear a trivial question, but in an era where adults interacting with little children is considered suspect behavior until proven otherwise, it is not.

Nonetheless, in the past or in most rural places, still anyone nearby could be a suffering child who would ask for help to another child or an adult in the safe web built around the child. The simple interaction could help to restore the natural condition of the ancestors: the child would be asked, for instance, "what's your name?" or "how do you feel?" and then a chain of events could be triggered. This kind of "natural network" can be transposed to a virtual level where subjects have their virtual counterpart.

7.5 Spread Surveillance

If all data are managed as stated up to now, there could be many public information globally available: even complex medical monitoring parameters could be made public, as long as the ownership stays unknown. A rare pathology could be managed storing periodically collected parameters on the Web, so that medical consultancy would be easier (no more tons of emails sent forth and back) if all involved doctors could have access to a common platform: moreover, a positive side effect – real-time free instant surveillance – could be obtained.

7.5.1 The Proposal: Real-Time Free Instant Surveillance

A further proposal can be therefore made: through an already work-in-progress Web service, medical data can be made *public*, but *anonymously* (if not set otherwise) available; doctors, researchers, and volunteers can get these data filtered in their competence area (for study, curiosity, work, research reasons).

In such an established model, anomalies can be detected by hundreds of observers and notifications sent to the "registered" user (maybe the owner and/or the doctor or, in the case of a child, to his caregiver).

It would help to the construction of a sustainable system in which real-time data can be examined (raw data that do not suffer of the limitations of ad hoc studies and consequently can fill those gaps that clinical trials or randomized studies could have, even if with all the restraints that improperly designed research brings inside) and benefits can be offered to those that decide to donate their personal data or information in an anonymous way or not.

Conclusions

In an ever-growing community of Web-connected people, cooperation and interaction give rise to new opportunities: not wasting them is a reachable effort. The number of data produced every day and the net of connections that is established hour by hour are a tremendous source of resource that can be exploited in behalf of children's health.

The rising generations already experience in their routine how the digital era is operating, with the youngest one more confident with electronic devices rather than traditional communication tools. Most of their communication already happens today through very easy-to-use, cheap, accessible, and intuitive instruments that are a prominent part of their life.

This also applies, with a number of limitations, to developing areas and in very rural districts of lowest-income countries.

The implementation of telehealth models has to take into account this scenario, exploiting the already operating resources (whose technological level is already very high and already able to support most of the chores needed by a telemedicine system) in order to obtain instruments children could become familiar with, so that they can be easily used and "promoted": from everyday ordinary tools to advanced health management devices.

Major issues related to cost-effectiveness, power supply, privacy, or liability have to be properly addressed, provided that children are already partially aware of some of those questions (sharing or browsing for inappropriate contents, buying devices with a better battery capacity, and so on) and that the Web-based software and hardware available today – and according to the model presented, usable for telehealth purposes – have already dealt with those problems, starting from their design and implementation.

References

1. VVAA. Leonardo da Vinci. Wikipedia, the free encyclopedia. Available from: http://en.wikipedia.org/wiki/Leonardo_da_Vinci. Accessed 11 Mar 2014
2. Kambayashi Y, Mohania M, Min Tjoa AM (2000) Data warehousing and knowledge discovery. Springer, London. p 49. ISBN 9783540679806
3. Yantzi N, Rosenberg MW, Sharon O, Burke S, Harrison MB (2001) The impacts of distance to hospital on families with a child with a chronic condition. Soc Sci Med 52(12):1777–1791
4. Probst JC, Laditka SB, Wang JY, Johnson AO (2007) Effects of residence and race on burden of travel for care: cross sectional analysis of the 2001 US National Household Travel Survey. BMC Health Serv Res 7:40
5. Karp WB, Grigsby RK, McSwiggan-Hardin M, Pursley-Crotteau S, Adams LN, Bell W, Stachura ME, Kanto WP (2000) Use of telemedicine for children with special health care needs. Pediatrics 105(4):843–847
6. Rinaldi G, Gaddi AV, Capello F (2013) Medical data, information economy and federative networks: the concepts underlying the comprehensive electronic clinical record framework. Nova Science Publishers Inc, Hauppauge. ISBN 978-1-62257-845-0
7. The Economist. Data, data everywhere. Special report: managing information. Available from: http://www.economist.com/node/15557443. Accessed 11 Mar 2014

Part III

Complex Scenarios and Special Settings

8 Rural and Extreme Rural Settings: Reducing Distances and Managing Extreme Scenarios

Fabio Capello

A major and somehow logical application for telemedicine is its implementations in those areas that because of geographical reasons are far from health centers and hospitals. Millions of people around the world today, both in high- and low-income countries, cannot easily access any medical services because of the distance. The use of connecting devices and point-of-care analyzers, able to replicate a conventional doctor-patient encounter, is a first possible solution. On the other hand, extreme conditions in very rural settings can create restraints to the use and the maintenance of delicate machines.

When it comes to children, some other considerations have to be made. Firstly, the number of children that today inhabits regions with extreme conditions is quietly rare in high-income countries, while it can be considered a standard condition in very poor areas [1–4]. Secondly, the implementation of sophisticated medical equipment goes together with the progress of an area, accordingly to its technological development [5, 6]. There is no telemedicine if there are no available connections. Those same connections, because medicine cannot depend on delays or given timesheets, have to be continuously reliable and available 24/7.

There are many possible ways to achieve such a connection: it can be wired, wireless, mobile, and satellite based. Each one of these has a cost and limitations. It is not the main aim of this book to discuss restraints and opportunities related on a technical discussion, yet some considerations able to explain why so many projects in telemedicine for rural areas and developing countries fail have to be made.

Because of that, two main categories have to be considered: those telemedicine strategies to be used in remote areas, rural settings, and areas with extreme conditions for high-income countries (including those areas that do not lie properly inside the geographical boundaries of a country as oil platforms, isolated research stations like the ones in Antarctica, or space stations but that could make use of very

F. Capello, MD, MSc
Pediatrics and Child Malnutrition, CUAMM – Doctors with Africa,
Via S. Francesco, Padova, Italy
e-mail: info@fabiocapello.net

F. Capello et al. (eds.), *Telemedicine for Children's Health*, TELe-Health,
DOI 10.1007/978-3-319-06489-5_8, © Springer International Publishing Switzerland 2014

developed health systems and high health budgets) and remote and rural settings in developing countries.

As we have seen, in the first ones, children cannot be considered the main users of the health system, whereas in the latter, underage people often are.

This is a key point, in the development of telehealth systems for children that live in those areas, also related to the fact that the resources those different cadres of people can count on are very different, the main consequence being the implementation of personal devices for telemedicine versus social-oriented devices available in public health centers.

A singular scenario is the one of the big emergencies and catastrophes, in which these two different settings often match.

8.1 Telemedicine for Rural Areas, in High-Income Settings

In 1955 the World Health Organization released a monograph related to the organization of a rural hospital in which it was stated that "the rural health system must be flexible and energetic in action, and must be in closest possible touch with an urban centre, well-equipped in staff and material. The question of communication is consequently one of cardinal importance [7]."

Nowadays, things are not different. Communication is still a crucial issue in the rural scenario, as the goal is to create a connection among rural and urban health centers, in order to virtually dissolve the distances.

Two main features characterize the delivery of health in those scenarios. The first one is the widespread availability of high-quality and high-speed Internet connections that go together with the high diffusion of ICT devices. In most industrialized countries, connecting devices today are cheap, reliable, easy to find, and easy to service. Almost every family has at least one mobile phone or a smartphone and a personal computer, with many of the families with at least one of these devices for each component of the family, children included [8, 9]. It means that it is common to find young people, also under the teen and the tween years, that own their personal connecting device. Besides, children in those settings are very really acquainted with those technologies [10], most of them being able to use and understand them even at very young ages (below 2 years of age) [11].

The second main feature is the organization of the national health systems of high-income nations. This is not strictly related to the quality of the service of the accessibility to the same service for everyone. Yet in industrialized countries, the health systems can generally count on very high budgets or on private services that could handle very sophisticated technologies and trained and highly motivated personnel. In addition, there are no chronic shortages of health workers or supplies, at least compared to those developing areas in which the patient/doctor rate and the health worker/patient rate are often unfavorable.

Because of that, the construction of telemedicine systems able to follow those criteria that have been analyzed in other chapters of this book (see Part I) could be possible, and the modifications needed to adapt the telehealth models to those settings are theoretically relatively few.

Nevertheless, there are many physical and geographical restraints that have to be considered and properly addressed (e.g., transportation, spread of population over the territory and distances among households, assistance and maintenance of the installed devices and therefore the availability of technicians able to travel to the remote area in case of need, coverage and power supply of relay stations, limitation of actions due to the environmental conditions, and effect of environmental conditions on sensors and equipment). Moreover, in those cases in which telemedicine cannot cover the needs of populations or of selected patients, users cannot simply go from remote areas to a hospital or to a health center to receive standard care. Therefore, a telemedicine model designed for those scenarios has to consider the opportunity to build systems to deliver traditional care also in very extreme areas.

Tele-surgery that exploits robotics can be an example for all. This is a very expensive technology, already in use in some centers, that allows distant operators to work on a patient, tele-guiding a robot that could work in their place. This means that different professionals that cover different areas of expertise can theoretically operate on a patient, being the alternative the residency of a team of surgeons from different specialties in the same remote area where one single robot has been installed.

This is crucial when it comes to children: a number of conditions that could affect children are rare, and the opportunity to send several specialists that could cover all those fields of medicine, in areas in which there are relatively few children and in which the chance to have a child with a selected disease is poor, basically made no sense. There would be no point, for instance, in sending a child doctor, specialist in cystic fibrosis or in epilepsy, in an area with a very low number of children, just because someday a child with one of those illnesses could show up.

Some authors suggest the use of Web portals that could directly link patients seeking for help with doctors and specialists in order to reduce the distances, also reducing the need for the installation of specific equipment in a physical health center [12]. This can be an interesting option that needs to be explored, also enhanced by the tools that came from the Web 2.0 or from virtual reality technologies that are likely to be implemented in the future. This may help to recreate a virtual surgery in which patients and health professionals could enter in touch miming the traditional doctor-patient encounter.

8.2 Extreme Scenarios

8.2.1 Major Emergencies, Natural Catastrophes, and Humanitarian Scenarios

The delivery of health has to deal with situations that could hit the system in every moment. This is particularly true for every emergency service. Because of their inner nature, disasters and major accidents represent situations that cannot be foreseen (or only partially expected, as in the case of armed conflicts, wars, and famine) and that can happen anytime and in any place. Natural catastrophes and humanitarian scenarios can hit any population in spite of their economic and social background.

In such occasions, children are probably the most vulnerable casualties. Besides the events that create the crisis, in fact, children can suffer more than others of famine, epidemic outbreak, or neglect secondary to the catastrophe itself. Posttraumatic stress, besides, can easily affect the youngest, increasing the distress and the need for help and support.

In those situations, the background is not different from the one of extreme rural and remote settings, in which there is a limited availability of medical resources and reduced patient/doctor rate and where medical facilities can become completely cut off from the disaster area (in the abstract, those are scenarios that can overlap with the ones described in Sect. 8.3). In addition, though, the emotional impact of the disaster, the possible existence of collateral risks, or the perseveration of the same condition that creates the humanitarian emergency (an earthquake or a flood that resolves itself after the main event, with or without collateral damages as fires or collapses of damaged buildings versus a war conflict that spreads across a region getting worse and affecting more people and areas over time) can increase the level of complexity of the setting to handle.

The use of ICT devices and Web-based systems and user-generated data has already been proved useful in some major emergency [13] and has been also fostered by the United Nations and OCHA [14]. In major disasters, civil defense and fire brigades already ask private users to remove the password to their Wi-Fi connections in order to give access to the Internet for those who are seeking for help [15].

Two main fields of application can be defined for telemedicine in these occasions:

- *Organization, optimization, and management of the health and human resources in the field*: Emergency scenarios have to deal with chaotic conditions that are part of the same nature of the disaster itself. Most of the real circumstances that take place cannot be controlled following a given plan, and some situations require a real-time upgrade of the crisis management. Yet some of the main features can be predicted and organized according to accurate schemes. Algorithms of intervention and real-time upgrades of the same plan of intervention are major components of the approach to great emergencies, especially when working with children, whose fragile conditions can vary very rapidly over time. A control room for the management of the intervention and of the human resource in the field does not have to be in the same place hit by the emergency. On the contrary, it would become a limiting factor, as the same coordinators would be possibly subjected to the same hostile conditions people they are caring for are affected of.
- *Teleconsultation*: Because of the restricted number of workers that can operate or that can reach the area, the need for teleconsultation can be crucial, especially when dealing with children. It is common sense that task forces or rescue teams can or cannot have personnel especially trained to work with children. Besides, even in the best possible scenario, not every field can be covered by the people that reach the field. The use of ICT tools can help health workers to receive instruction in real time from specialists from around the world (e.g., a radiologist that could read an x-ray or a US scan in remote or an expert in infectious diseases that could help to deal with outbreaks). In those special situations in which medical doctors cannot reach the area, health workers, nurses, or paramedics can be authorized to

perform procedures and to administrate treatment under the guide of child doctors in remote. Point-of-care devices and portable analyzers can be used to assess specific patients and to collect and send in remote health data of single patients or of a cluster of patients. This can help to improve the care of single patients and to state the health condition of the community in the disaster area.

Major concerns related to the use of ICT devices in extreme scenario (see also Chap. 10) can be the environmental conditions that could affect those same tools, compromising their functionality. This is because electronic equipment can be extremely delicate or breakable, but also because they require continuous power supply and a reliable connection. Yet power supply, Internet, and telephone (wired or mobile) connections are particularly at risk in case of major accidents.

A plan for action that aims to use telemedicine systems for the management of the intervention and for the delivery of the case needs an efficient logistic able to provide power and connection at any time. Different strategies can be used (i.e., solar panels, power generators, satellite connections), and those mainly depend on the area of intervention and the resource both available in the area and of the one owned by the rescuing teams (or that can be easily dispatched to the area).

There would be no point, for instance, in using instrumentation supplied by solar energy in a polar region in winter time.

8.2.2 Extreme Rural Areas in Developed Countries

As we have seen, extreme rural areas exist also in very high-income countries in which national health systems or private health systems can provide a first-rate healthcare or in which the availability of resources for health is not a limiting factor. In those same settings, technology applied to health is largely available and based on the ultimate systems designed for care.

Because of that, the availability of medical equipment also in remote areas should not be virtually considered an issue. On the other hand, in areas with a very low population density (as the northern territories of Canada or the inner region of Australia), it would make no sense to install very expensive devices that could serve a very limited number of children and sometimes for very rare medical conditions [16].

Yet because such situations – also with families that live in an isolated area with children – exist, solutions have to be found [17]. Flying doctor services are currently operating in many regions of the world (as the Royal Flying Doctor Service in Australia or the Flying Doctors of Canada). This is a possible solution in which highly trained professionals bring their competences to a specific area. On the other side, doctors with a wide area of expertise rarely are specialists in a specific field, or even if it is the case, they cannot be specialists in every field.

Two main areas of intervention can exploit telehealth technologies in such situations:

- *Teleconsultation*: Medical doctors that move from an area to another can ask and receive in real-time consultation from different specialists that assist and support the professional in the field.

- *Use of portable devices and analyzers* that can check up the state of health of the patient, sending data to remote control centers. Expensive and sophisticated devices (which also need trained personal to be used) in this case do not have to be installed in the given area, but can travel with the mobile service. This has to be considered as a decisional assistance for those professionals that work in the field in extreme scenarios, aimed also to support a medical consultation made by a specific specialist in order to solve specific conditions.

In addition, applications that exploit personal communication devices, as mobile telephones, personal computers, or smartphones, have to be considered. This approach is not different from the connectivity network for telehealth that can be implemented in the community of urban areas or less extreme rural areas (see also Chap. 3).

In those areas in which basic health centers exist, the installation of permanent devices, based on systems designed for specific and general medical purposes, also has to be considered.

The availability of power supply and of 24/7 connections and the existence of a hostile environmental condition that could affect the function of the medical devices needed for telemedicine are restraints that have to be carefully considered on a case-by-case basis.

8.3 Future Scenarios

Telemedicine can be considered a technology available today. Especially, based on personal connecting devices, there are many applications that are already in use or under development. Others are foreseeable, hopefully as part of a proper eHealth theory, and able to follow the models hereby presented.

Yet telemedicine can still be considered an instrument of the future.

The application of these technologies has to take into account those scenarios that by now can be considered science fiction. When compared to the wireless world (meaning not connected at all) of the beginning of the twentieth century yet, the world of today could appear as futuristic as a science-fiction novel.

It implies that in the future children could live in those same settings in which children have no role today, as space centers or underwater worlds. Those extreme examples show how some situations can be completely detached from the ordinary world. In those scenarios, the self-sufficiency of a health system would become crucial. No aid in fact could be asked from people living in a completely isolated world.

This is already a big issue today that space medicine has to deal with [18–20]. But manned missions to outer space, for instance, are likely to have onboard families in the future, especially for longer ranges.

Those are extreme examples, but they focus on the fact that in the future, the systems for telemedicine should become self-sufficient and able to operate without the direct intervention of outer operators.

The efforts of developers, stakeholders, and enterprises have to consider this ultimate goal, keeping always in mind that just because a child would not be in the same room with a doctor or with health workers, it does not mean that the child is not a person like the ones that doctors meet every day in their professional life.

This is crucial. Otherwise, the development of those systems – and especially the ones thought to operate among completely detached settings – could produce a risky depersonalization of the same medicine. It would be like pushing a button, knowing that this action would kill a child in another part of the world and – not knowing or having direct contact with the kid – living easily and getting away with that.

Again, this is an extreme instance, but it would to help to understand those issues that today telemedicine for children already has to face.

Conclusions

Because of the wide spread of mobile and Web-based technology, the use of telemedicine for children living in rural settings can be considered a feasible and sustainable solution for the delivery of care in those areas with a very low population density and that are generally cut off from the main health services [21]. As long as telemedicine service has to be theoretically available 24/7, the application of those models has to be planned according to the peculiarities of the settings in which it is supposed to take place, considering all the possible restraints that could limit the use of the system.

Major emergencies, in which children can be considered especially exposed and that have several features in common with other extreme scenarios, could greatly exploit the opportunities that come from ICT tools, provided that correct planning is granted.

Future scenarios have to be considered, in order to prepare telehealth models that could fit also those contexts that still belong to tomorrow's people, but that nevertheless are due to become the daily nature of the world still to come.

References

1. Shackman G, Wang X, Liu YL (2012) Brief review of world demographic trends – trends in age distributions. Available at SSRN: http://ssrn.com/abstract=2180600 or http://dx.doi.org/10.2139/ssrn.2180600. Accessed 10 Mar 2014
2. Population Distribution of Children and Youth in the Countries ChildFund Serves. United Nations Population Program and United Nations Development Program Data for 2013. http://www.childfund.org/uploadedFiles/public_site/media/articles/current/2013/Population%20Distribution%20of%20Children%20and%20Youth%20table.pdf
3. Government of Canada. The Canadian population in 2011: age and sex. Canada Statistics. Catalogue no. 98-311-X2011001. ISBN 978-1-100-20670-7
4. Lichter DT et al (1979) Trends in the selectivity of migration between metropolitan and nonmetropolitan areas: 1955–1975. Rural Sociol 44(4):645–666
5. Hidalgo CA (2010) Graphical statistical methods for the representation of the human development index and its components. United Nations Development Programme Human Development Reports. Human Development Research Paper 2010/39
6. Castells M (1999) Information technology, globalization and social development. United Nations Research Institute for Social Development. UNRISD Discussion Paper No. 114. Geneva
7. Bridgman RF (1955) The rural hospital. In: Structure and organization. World Health Organization Monograph Series No. 21. Geneva

8. Lenhart A, Ling R, Campbell S, Purcell K (2010) Teens and mobile phones. Pew Research Center. Available from: http://www.pewinternet.org/2010/04/20/teens-and-mobile-phones/. Accessed 10 Mar 2014
9. Kalba K (2008) The global adoption and diffusion of mobile phones. Harvard University – Center for Information Policy Research. ISBN 0-9798243-0-3 P-08-2. New Haven
10. Olson KE, O'Brien MA, Rogers WA, Charness N (2011) Diffusion of technology: frequency of use for younger and older adults. Ageing Int 36(1):123–145
11. Ward V (2013) Toddlers becoming so addicted to iPads they require therapy. The Telegraph. 21 Apr 2013. Available from: http://www.telegraph.co.uk/technology/10008707/Toddlers-becoming-so-addicted-to-iPads-they-require-therapy.html. Accessed 10 Mar 2014
12. Labiris G, Coertzen I, Katsikas A, Karydis A, Petounis A (2002) An eight-year study of internet-based remote medical counselling. J Telemed Telecare 8(4):222–225
13. Hern A (2013) Online volunteers map Philippines after typhoon Haiyan – Humanitarian OpenStreetMap Team coordinates mapping effort after enormous storm devastated country. theguardian.com, Friday 15 Nov 2013. Available from: http://www.theguardian.com/technology/2013/nov/15/online-volunteers-map-philippines-after-typhoon-haiyan. Accessed 10 Mar 2014
14. Gilman D, Noyes A (2012) Humanitarianism in the network age. OCHA – United Nations Office for the Coordination of Humanitarian Affairs. United Nations publication. Sales No. E.13.III.M.1 ISBN-13 978-92-1-132037-4
15. Caffo A (2012) [How to open Wi-Fi to help victims of the earthquake]. Art. In Italian. La Stampa, 30 May 2012. Available from: http://www.lastampa.it/2012/05/30/tecnologia/come-aprire-la-connessione-wifi-per-aiutare-le-vittime-del-terremoto-vU2dZ0UAQ3SabIF4SiFRhM/pagina.html?exp=1. Accessed 10 Mar 2014
16. Gruen RL, Weeramanthri TS, Knight SSE, Bailie RS (2004) Specialist outreach clinics in primary care and rural hospital settings. Cochrane Database Syst Rev (1):CD003798
17. Smith D (2000) Critical care at the electronic frontier of the 21st century: report from the 29th educational and scientific symposium of the Society of Critical Care Medicine, Orlando, USA, 11–15 February 2000. Crit Care 4(2):101–103
18. Orlov OI, Grigoriev AI. Application of telemedicine technologies to long term spaceflight support. IAF abstracts, 34th COSPAR Scientific Assembly, The Second World Space Congress, Houston, 10–19 Oct 2002, p G-5-03, meeting abstract. Available from: http://adsabs.harvard.edu/abs/2002iaf..confE.178O. Accessed 10 Mar 2014
19. Cermack M (2006) Monitoring and telemedicine support in remote environments and in human space flight. Br J Anaesth 97(1):107–114
20. ESA – European Space Agency (2012) Innovation for health. www.esa.int. 13 Nov 2012. Available from: http://www.esa.int/Our_Activities/Technology/Innovation_for_health. Accessed 10 Mar 2014
21. Dharmar M, Romano PS, Kuppermann N, Nesbitt TS, Cole SL, Andrada ER, Vance C, Harvey DJ, Marcin JP (2013) Impact of critical care telemedicine consultations on children in rural emergency departments. Crit Care Med 41(10):2388–2395

9 Telepediatrics in Developing Countries: A Better Care for Children in Low-Income Settings

Fabio Capello

The use of advanced communication technology in the low-income setting has been presented as a possible solution able to cope with the lack of medical and human resource in developing countries. Yet this assumption has a paradox in itself: how can a developing area (technologically underdeveloped and with limited resource) adopt an advanced and possibly expensive model? This is not only a theoretical or logical fallacy. In fact, any project aimed to implement a health system or interact with an existing one has to deal with three basic questions:

(a) *The assessment of the need in the area where the project has to be implemented:* There is no point in building a hospital or a health facility in the middle of the Sahara desert as long as no patient is apt to travel for miles under the sun and possibly die in the attempt, in order to receive assistance. This is common sense, but there are also anthropological, ethical, social, cultural, religious, political, and economic implications that have to be addressed and taken into account.

(b) *Capacity issues*: In some cases, those are binding statements and do not allow the implementation of new programs. Because the advanced communication system implementation in a developing area often needs the involvement of outer stakeholders, foreign enterprises or governments have to follow the rules of the country they want to assist. But in many cases, strategic plans of those outer parties collapse with the ones implemented by the hosting country. This could sound unreasonable because of the higher expertise of a most technologically advanced partner. On the contrary, because the outer stakeholder often belongs to a different world, most of the projects it could imagine simply do not fit into the reality they would like to work in. The use of electronic systems, for instance, needs a reliable and continuous power source. Yet many rural areas still do not have electric energy. Many deeds that could be taken for granted (as recharging the battery of a mobile phone or plugging a PC) simply

F. Capello, MD, MSc
Pediatrics and Child Malnutrition, CUAMM – Doctors with Africa,
Via S. Francesco, Padova, Italy
e-mail: info@fabiocapello.net

F. Capello et al. (eds.), *Telemedicine for Children's Health*, TELe-Health,
DOI 10.1007/978-3-319-06489-5_9,

cannot be done. Another major issue is the lack of technical assistance that medical and high-technology equipment constantly need, for maintenance or calibration. We would not discuss in this book the main technical, geographical, political, and anthropological issues that limit the implementation of advanced devices in underdeveloped areas, but it has to be strongly stated that an in-depth analysis of any single area in which a program of telemedicine has to be implemented is crucial.

(c) *The assessment of local capacity*: Both in terms of physical resources (expendable budgets, possibility of installing a maintain a high-technology equipment, and so on) and human resources (in terms of numbers and capacity of using those resources). What the so-called capacity building is, is a key issue for the empowerment of the least advanced countries. This is especially true in healthcare. The correct evaluation of the resource that can be exploited is crucial. No program can be implemented if there are no means to execute and above all if it cannot be sustainable over time. Telecommunication needs great investments since the beginning (direct cost of equipment) but also needs constant maintenance and a continuous supply of disposable or consumable items. Moreover, advanced technologies need trained personnel able to use them. The human and economic resources are the bricks a health program has to be built with. Any project that requires a constant input of funds or workers that come from the outside is bound to fail if not planned according to the concept of sustainability and local ownership that are strictly connected to the idea of capacity building.

In addition, it has to be stated that every single country has its own peculiarities. Many projects implemented so far have been planned according to general indications that come from literature and that assess the need and the restraints in generally defined "developing countries." Every country, or better, every single region, has a proper unicity that depends on the culture and the tradition of each ethnic group and that cannot be mismatched with the ones of its neighbor tribe. The cultural boundaries and the differences among the different people and countries are so strong that what can be appropriate in a particular area could be outrageous some fifty miles away. In some cultures, for instance, people ask continuously for photos: wearing a camera around means stopping every few steps to take a picture. For some other ethnic group, photography is a taboo. In other regions, mainly in metropolitan areas, secondary to legal issues (people are afraid that a picture can be taken to a police station), people do not want to be photographed and therefore possibly be identified.

This example that is also useful considering the possible applications of telemedicine – which often require for pictures to be taken – helps to explain how cultural difference among two populations can impede the development of a project. This is the way any intervention should be tailored on the population, starting as we have seen from the perceived and real needs and from the availability of resources in the setting.

Nonetheless, the increasing speed with which communication technology is spreading around the world, also in settings with very limited resources, can help to overcome most of those restraints, provided that the correct analysis of those features that influence this phenomenon is considered, so that adequate plans could be designed [1].

9.1 Creating a Network in Behalf of Children

One of the main peculiarities of the developing areas is the high birth rate and consequently the percentage of underage people that make up the population (see Tables 9.1, 9.2, and 9.3) [2, 3]. Many are the reasons that lie underneath (see Table 9.4), and most of them are due to change once the economic and the cultural backgrounds change.

Table 9.1 The number of children under the age of 5 and under the age of 18 that live in the least developed countries

	Population[a] 2012, total	Population[a] 2012, under 18	Population[a] 2012, under 5
Least developed countries	878,097.5	412,086 [46.93 %]	130,370.3 [14.85 %]

[a]Thousands

Table 9.2 The number of children under the age of 5 and under the age of 18 that live in the most industrialized countries

	Population[a] 2012, total	Population[a] 2012, under 18	Population[a] 2012, under 5
Japan	127,249.7	20,310.4 [15.96 %]	5,389.4 [4.24 %]
USA	317,505.3	75,320.5 [23.72 %]	20,623.4 [6.50 %]
Western Europe	414,561.80	78,714.00 [18.99 %]	22,160.90 [5.35 %]
Australia, New Zealand, UK	90,293.5	19,677.1 [21.79 %]	5,888.3 [6.52 %]

[a]Thousands

Table 9.3 The number of children under the age of 5 and under the age of 18 that live in the world

	Population[a] 2012, total	Population[a] 2012, under 18	Population[a] 2012, under 5
World	7,040,823.1	2,213,677.1 [31.44 %]	652,093.4 [9.26 %]

Source: UNICEF
[a]Thousands

Table 9.4 Factors that affect birth rates and fertility rate worldwide [4]

Children as a part of the labor force. Rates tend to be higher in developing countries (especially in rural areas, where children begin working to help raise crops at an early age)
Urbanization (see Table 9.7). People living in urban areas tend to have fewer children than those living in rural areas, where children are needed to perform essential tasks
Cost of raising and educating children. Raising children is much more costly in developed countries (more unessential goods purchased; children enter the labor force later)
Educational and employment opportunities for women
Infant mortality rate. In areas with low infant mortality rates, people have fewer children because fewer children die young
Average age at marriage and of the first pregnancy. Lower in least developed countries
Availability of private and public pension systems. No need for many children to help support parents in old age
Availability of legal abortions
Availability of reliable birth control methods
Religious beliefs, traditions, and cultural norms

Table 9.5 The level of mortality under the age of 5 (2012)

	Under-5 mortality rate	Infant mortality rate (under 1)	Neonatal mortality rate	Annual no. of births[a]	Annual no. of under-5 deaths[a]	Life expectancy at birth (years)
USA	7	6	4	4,225.7	29	78.8
Western Europe	4.05	3.25	2.35	188.67	0.85	81.50
Australia, New Zealand, UK	5.53	4.63	3.18	188.67	14.93	81.50
Least developed countries	85	58	30	29,286.7	2,388	61.1
World	48	35	21	138,313.8	6,553	70.6

[a]Thousands

Table 9.6 The level of literacy when compared between Western countries and least developed countries

	Total adult literacy rate[a] (%)	Primary school net enrolment ratio[b] (%)
USA	–	95.7
Western Europe	99.0	97.50
Australia, New Zealand, UK	–	97.50
Least developed countries	58.5	80.7
World	84.1	91.2

[a]2008–2012
[b]2008–2011

Table 9.7 The percentage of people living in urban areas in different parts of the world (2012)

	Urbanized population (%)
Western Europe	76.1
Australia	89.4
New Zealand	86.3
Japan	91.9
USA	82.6
Least developed countries	28.9
World	52.5

Nonetheless, the life expectancy (the mean number of years that a child born today is due to live in a different part of the world) varies according to the place where the child was born, with a peak in the most industrialized countries and a very low average in those countries with poor incomes and rate of school attendance and education (see Tables 9.5 and 9.6) or urbanization level (Table 9.7).

This incredible difference among the different countries should not be considered bearable, and most of the efforts made by the international community and by the government and nongovernment organizations – although with many flaws – are directed to the improvement of the children's health worldwide.

Infectious disease and perinatal complications (asphyxia, sepsis, congenital, prematurity, etc.) are among the main causes of death or morbidity in children belonging to underdeveloped areas [5]. Moreover, in one third of all children deaths under 5 years, malnutrition is a significant cofactor [6].

Those are often curable diseases that do not need sophisticated diagnostic tools or a complex therapeutic approach. On the other hand, the management of a child is difficult and requires a specific training that could help the health professional cope with the peculiarities of the youngest ones.

Even when the medical training and the education of health workers focus also on childhood in those settings, the lack of professionals with a specific and specialist training is a current issue.

This might lead to a poor management of the ill child, with some of the patients killed or impaired secondary to the treatment that they had received in the hospitals or in the health centers.

This opens two main issues that telehealth is asked to address:

- The education of the personnel working mainly or selectively with children (see also Chap. 11)
- The support to those health workers and caregivers for the management of an ill child (prevention, diagnosis, management, and follow-up of acute, subacute, or chronic conditions)

Distant learning and eLearning programs – even with some limitations – are under development or already implemented that could help to give an answer to the first question (see also Chap. 11).

The use of ICT tools for the management of a child's health can exploit two main features of telehealth:

- The possibility to give support and medical advices to those health workers that have to deal with children, but that do not have a specific training. This is due to improve the reliability of a diagnosis and help to deliver the correct treatment to the right patient.
- Monitoring from the distance the state of health of a population – focusing on children – so that specific programs of intervention can be undertaken, organized, and managed.

Continuous feedbacks should monitor the outcomes so that adjustments and reprogramming can be made, tailoring each project on specific situations.

9.1.1 Management of Single Cases

The medical practice in rural areas of very low-income countries has several restraints that have to be considered. This was true once [7], but still applies to many scenarios around the world today, where the level of development is still comparable to the one of the past. Because of shortage of health staff, a fine investigation and management of the single cases are generally not possible. In a mean rural hospital, the quality of care is relatively lower compared to the ones in urban areas [8]. Specialist outreach can become impossible. Often the same specialist has to act as a general practitioner or has to take care of patients he or she has not the ability to care for. This is particularly true for pediatric cases, as long as many pediatric wards and consultations in disadvantaged hospitals are under the responsibility of physicians that never had specific training in children's health. Moreover, in most settings there is no medical doctor, and rounds and medical decisions are under the responsibility of trained nurses.

In any case, often the availability of therapeutic choices is limited, and the medical doctor has very few drugs he can use. That basically makes the need for fine diagnosis useless.

In addition, the high number of patients attended in the OPDs or that have been admitted to children's wards reduces the possibility for doctors and nurses – especially when the human resources are limited – to dedicate enough time to specific patients.

Nevertheless, a number of cases need an in-depth investigation all the same.

Some possible very easy solutions can be already used to enhance the possibility of diagnosis and of therapeutic success in those selected cases. Basic email or telephone consultations – both based on encounter scheduled on regular basis or on ad hoc consultation for specific cases – have already been proved effective in some occasion [9]. Mobile phones can take a picture and send short messages or emails to distant consultation centers in real time [10]. Web-based portals can be designed to assist medical doctors working in rural areas of developing countries, so that real-time specialist consultation could be conducted. The use of point-of-care devices also has to be considered (see also Chap. 9), although in the poorest areas, their implementation can be undermined by hostile environmental conditions, especially in critical scenarios or major emergencies [11].

On the other hand, it is crucial to keep those consultations only for very selected cases: as we have seen, the load of work in rural areas in developing countries is often very high, the therapeutic choices are limited, and the time needed to ask for an e-consultation is time detracted to other children that could need immediate assistance.

Besides, parents are not always keen to wait for the response of a possible teleconsultation, as long as attending a hospital means leaving the rest of the family (siblings of the ill child, elderly people, children that are somehow related to the family) and their everyday duties apart. This is not a minor issue, as long as many of those families can count only on their own work (farm their own land, keep the animals, collect water) to survive.

Nevertheless, children are generally prone to acute conditions that require standard treatment.

Thus, teleconsultation has to be encouraged provided that very basic resources are needed and used.

On the other hand, technology is rapidly changing, and very cheap connection systems will be probably available in the near future. Children's wards in rural hospitals and in poor settings have to become a detachment of urban hospitals. Virtual rounds, with a nurse operating on basic connection devices (tablet or smartphone, for instance) and a doctor prescribing medical orders on the other end, could easily take place. This is probably not going to improve the outcomes in itself if the available drugs are few, but it can help to create a continuous feedback among health workers in the rural area and health professionals in remote that is likely to enhance the commitment of the same health workers and the compliance to the treatment by children and families. Parents in fact often need to be reassured and guided by health professionals able to establish a difference among the cure that children receive in the hospital and the ones that they have from traditional medicine and witch doctors.

9.1.2 Teleconsultation in Urban Hospitals of Least Developed Areas [12]

Many of the issues discussed so far apply mainly for rural hospitals in very low-income or undeveloped areas. Nevertheless, those same problems can also be present in urban hospitals or in university clinics where the availability of medical resource and trained personnel is higher.

Telemedicine can play an important role, as long as the specialist outreach could become possible in those settings. Urban hospitals can generally count on a higher and more diversified number of health workers and specialists and on more organized schedules. The number of patients attended could become an issue, but some of them are referrals that come from rural areas and that have the time and funds to afford the longer waiting that elapses from the assessment of the patient to the reply of a teleconsultation. In addition, central hospitals could have more therapeutic choices so that fine diagnoses that correspond to fine treatments can make sense.

Nevertheless, many families cannot simply afford to go to the city for further investigation, and in many cases, the child is not properly attended because nurses and doctors are unable to understand the problem of the patients (different dialects spoken or inability to explain the complaints and the past medical history).

In addition, in many cultures, when compared to elderly people that are relatively few as long as the life expectancy is poor, children are considered expendable, as long as the mortality rate under the age of 5 is very high and the mean number of children for each woman is relatively high (see Table 9.5). It means that rural health centers have to select for referral children that require for further assistance to urban hospitals. It means a proper assessment of the family and of the child's conditions before any action could be taken.

A net of referrals, based on electronic systems, could improve the quality of this service.

On the other hand, this requires a special commitment of health workers both in rural and urban centers, plus coordination from those health support centers that are due to deliver the final teleconsultation to the urban hospital.

This leads to a very complex web of relationships, with different possible critical points. The construction of a telemedicine system without these premises therefore is not to be recommended. Teleconsultation on a personal basis, required by individual doctors for selected cases, is still to be preferred, unless and until the above condition is not matched.

9.1.3 Epidemiology and Monitoring of the State of Health of the Population of Children Belonging to Least Developed Areas

An interesting field of application is the use of ICT systems for the real-time monitoring of epidemiological conditions of a given population of children. Because many are the indicators that are suggestive of the onset of an outbreak in a population, several are the options of monitoring and intervention. Whereas in higher-income countries the monitoring of the clustering of mobile devices or of the Internet searches related to particular medical conditions could lead to the mapping of medical scenarios, in poorer areas that could or could not be the case. Children generally do not own personal mobile devices, and the poorest families may not search the Internet to understand a medical term. Nevertheless, doctors and health workers even in rural districts are able to send data, even in real time, and to perform basic epidemiological surveillances.

Secondary to the load of work that health professionals have to face in those settings, the gathering of medical data is often neglected.

However, the use of very simple apps for touch-screen-based devices can help to overcome this issue (for instance, apps for a tablet where the major clinical conditions are indicated by icons. Doctors can touch the corresponding icon after every consultation. The number of cases is recorded in the tablet, in which the power supply is granted by rechargeable batteries to possible blackouts. As soon as an Internet connection is available, the apps automatically send in remote the epidemiological data collected).

Because the diagnosis is uncertain most of the time, and because in many areas the prevalence of some medical condition is unknown, some disease can be underdiagnosed, while others (as malaria in endemic areas) are overdiagnosed and overtreated. Epidemiological information, collected thanks to distant communication devices, should have to record also those suspect of disease that generally cannot be part of the patient record and cannot be stated in the final diagnosis (unconfirmed diagnoses). This can be crucial for the plotting of a map that describes all the possible outbreaks in those settings that cannot perform etiological diagnoses.

This is particularly important in children, as long as most of the campaigns of vaccinations are based on those data.

9.2 A Virtual Community, Embracing Developed and Developing Areas

The widespread distribution of ICT device as the mobile phone, also in very rural and very underdeveloped areas, mirrors the need for communication of even the most isolated population. This is a strong point that in part conflicts with what is stated in the previous paragraph and that could allow the construction of a web that is not different from the community children belonging to a richer world live in (see Chap. 3).

If that is the case, the boundaries and the distances among children and health workers could virtually not exist.

This is a very utopic vision of the world and of the future, also because of the cultural restraints among children belonging to different places, the ones among children and adult next of kin or caregivers, and the ones among children and health professionals due to take care of their health.

Nevertheless, the spread of the social Web is slowly affecting also the less developed countries [13], and it is reasonable to think that in the near future, children from very poor areas could have access to the Internet and to the Web 2.0.

Children, besides, have a natural attitude toward technology, as very young toddlers using confidentially tablets and smartphones demonstrate every day. Kids around the world already use mobile phones and access the Internet on a regular basis [14].

It is true that in many realities, the struggle for food, together with the problem related to child labor and exploitation or child abuse and neglect, is the everyday condition of most of the children belonging to poor rural areas. It is also true that many children from those same settings can access basic education and are apt to meet other children already acquainted with those technologies, and knowledge spread easily among children: the ones who know often transfer and teach – directly or indirectly – also to others.

The first step for a better health in children, also belonging to those backgrounds, lies in their awakening. Children can become aware of what their rights are and what a healthy life should look like. This is not going to solve their problem at once. Yet, it can help to reduce the distances.

The diffusion of TV has already achieved some of those goals. Children could reflect themselves in the ones that populated the shows they follow. They can live adventures and share emotions, tasting the same life of their TV's heroes.

Secondly, the spread of a common language can be achieved thanks to TV shows and Internet broadcastings. In many countries, children and families do not speak the national language, that is, critical restraints that limit the access to education (also to health education and therefore the access to prevention and early detection and management of acute or chronic medical conditions) and to care. Besides, the common language is also made by nonverbal communication and by shared perspectives and ideals.

Although the big risk is the loss of those traditions that create the culture and the identity of a population, the development of shared visions is an unavoidable and probably mandatory checkpoint in the road toward the development of a nation. It is true that many things can be lost on the way, but it is also true that the achievement of a better health for children passes also through this painful step.

Connecting people, therefore, can be the key to the building of a virtual society in which the goal is clear and the efforts to reach them are common.

Conclusions

Children belonging to the least developed areas should have the same rights in terms of delivery of care and well-being as children their age around the world. The natural contradiction that lies in the fact that underdeveloped countries do not have a suitable level of technology needed for the implementation of telemedicine systems can already be overcome today, thanks to the wide spread of mobile phones and easily affordable connecting devices.

Many are the restraints that delay the use of those models for health in limited settings, but if properly addressed – secondary to a study of the anthropological, ethical, social, cultural, religious, political, and economic issues appropriate for every single region – those can be solved and new opportunities disclosed.

The gold standard, and the final aim, would be the building of a virtual community in which, in spite of the physical location the child lives in, the access of care can be granted and top-rated specialist assistance offered.

References

1. Bakay A, Okafor CE, Ujah NU (2010) Factors explaining ICT diffusion: case study of selected Latin American countries. Int J Adv ICT Emerg Reg 03(02):25–33
2. UNICEF (2014) Statistics and monitoring. Customized statistical tables. Available from: http://www.unicef.org/statistics/index_step1.php. Accessed 14 Mar 2014
3. UNICEF (2004) The state of the world's children 2005. The United Nations Children's Fund, New York. ISBN 92-806-3817-3
4. Hawken P (2014) The human population: size and distribution. Available from: http://www.geowords.org/ensci/05/05.htm. Accessed 28 Feb 2014
5. Bryce J, Boschi-Pinto C, Shibuya K, Black RE, The WHO Child Health Epidemiology Reference Group (2005) WHO estimates of the causes of death in children. Lancet 365(9465):1147–1152
6. United Nations Interagency Group for Child Mortality Estimation (2012) Levels and trends in child mortality. Report 2012. United Nations Children's Fund, New York
7. Bridgman RF (1955) The rural hospital. In: Structure and organization, World Health Organization Monograph Series No. 21. World Health Organization, Geneva
8. Gruen RL, Weeramanthri TS, Knight SE, Bailie RS (2004) Specialist outreach clinics in primary care and rural hospital settings. Cochrane Database Syst Rev (1):CD003798
9. Smith AC, Youngberry K, Mill J, Kimble R, Wootton R (2004) A review of three years experience using email and videoconferencing for the delivery of post-acute burns care to children in Queensland. Burns 30(3):248–252
10. Goost H, Witten J, Heck A, Hadizadeh DR, Weber O, Gräff I, Burger C, Montag M, Koerfer F, Kabir K (2012) Image and diagnosis quality of X-ray image transmission via cell phone camera: a project study evaluating quality and reliability. PLoS One 7(10):e43402. doi:10.1371/journal.pone.0043402, Epub 2012 Oct 17
11. Kost GJ, Curtis CM (2012) Optimizing global resiliency in public health, emergency response, and disaster medicine. Point Care 11(2):94–95
12. WHO Library Cataloguing-in-Publication Data (2009) Telemedicine: opportunities and developments in Member States: report on the second global survey on eHealth. World Health Organization, Geneva
13. Galperin H (2005) Wireless networks and rural development: opportunities for Latin America. Info Technol Int Dev 2(3):47–56
14. Lenhart A, Ling R, Campbell S, Purcell K (2014) Teens and mobile phones. Pew Internet & American Life Project. Available from: http://pewinternet.org/Reports/2010/Teens-and-Mobile-Phones.aspx. Accessed 14 Mar 2014

Part IV

e-Learning

10 eLearning: Distant Learning for Health Professionals That Work with Children

Fabio Capello and Andrea E. Naimoli

Working with children could be tricky. Children have their own way of communicating and dealing with people. They express their emotions and their feelings in peculiar ways and end up with odd behaviors that are commonly considered inappropriate for adult fellows (as dancing in a supermarket aisle, making funny faces on their reflecting face on the window of a subway train, humming lullabies out of tune in a waiting room, etc.). The way they establish a relationship with adults varies accordingly to many factors, such as the age of the child, the sex, the age and the sex of the counterpart, the personal behavior of the child, or the attitude of the adults they are trying to communicate with. Besides, some children are keen to establish a relationship with grown-ups whether they like it or not, while others are really self-conscious and rarely communicate with adults or strangers unless they have to.

Even if communication skills are useful or essential, when it comes to health professions, even the most sociable adult can find a critical limitation in talking with the most young. Dealing with children means to learn again how to think as an underage. This is heavy because most of our adult life, since the teen years, has been spent trying to break off our childish part, in the attempt to create ourselves as independent and mature beings.

Therefore, health professionals and health workers that would like to spend their working life with children have to get used to the verbal, nonverbal, physical, and emotional communication strategies that children commonly use. Furthermore, they have to understand what are the main channels of communication that children prefer, including those childish topics that adults normally ignore: for a child, there are no boundaries between what is true and what is not (in their fantasies, Santa

F. Capello, MD, MSc (✉)
Pediatrics and Child Malnutrition, CUAMM – Doctors with Africa,
Via S. Francesco, Padova, Italy
e-mail: info@fabiocapello.net

A.E. Naimoli
Tech Department, Airpim Inc., Wilmington, DE, USA
e-mail: andrea.naimoli@elementica.com

F. Capello et al. (eds.), *Telemedicine for Children's Health*, TELe-Health,
DOI 10.1007/978-3-319-06489-5_10, © Springer International Publishing Switzerland 2014

Claus can actually travel one night a year all over the world, visiting all children's houses before dawn). This is part of the "magical thinking" already described by cognitive psychologists and that is a fundamental part of a child's intellectual development [1].

Children, moreover, live their illness in different ways: they feel different, when they compare themselves with other children free of their own disease [2]; they are keen to believe that they have been punished for something wrong they did, which also reflected in the feeling parents have toward those same children [3].

In addition, because of a number of problems related to children protection, privacy, or abuse and neglect prevention, those workers that work with children have to be specifically selected. This is becoming a major issue in Western countries, whereas other societies are less involved in those concerns. The right compromise among these two extreme standpoints has still to be established. Nevertheless, and whether it is extreme or not, parents ask for more guarantees for those people that are in strict contact with their children. Policies have to build also around those concerns, and institutes that offer training for people who work with children and that release qualifications and certifications of good standing have to consider carefully those issues.

Consequently, people who want to work with children have to go through specific trainings, both to develop and possibly exploit their inner abilities to communicate with the youngest and to become aware of the problems that are specific of younger ages (including data protection and privacy and prevention or early detection of mischievous behaviors against children).

To build capacity, especially in those areas where the access to education is poor, distant learning can be a possible solution.

10.1 eLearning

The discussion over eLearning is a long-standing one. Many have been the attempts to build a possible platform for distant education or for the enhancement of the traditional methods of teaching. Many universities already have Internet-based programs, and thousands of online courses are delivered every day worldwide, with more than seven million online students solely in the USA in 2012 with a growing rate of 6 % [4]. In addition, the electronic resources that vary from multimedia contents uploaded on social websites, user-generated contents, online libraries, or massive online open course (MOOC) made available much information targeted for different audience.

The basic idea lies on the concept of intercommunication: those who deliver the message and the learners have to get in touch. There are several strategies developed for this aim, while the media used to deliver the message can vary from TV or radiobroadcasts to computer-based and Internet-based applications up to mobile technologies (SMS and MMS on standard mobile phones, apps for smartphones and tablets). The first step is to establish which technology for whom. Besides, the feasibility and the sustainability of the projects over time have to be taken into account. This is particularly true when low-income settings are considered.

The main characteristics of distant learning are:

- *Time of delivery of the course*: The lecture can be given in real time, simulating frontal teaching, with or without an active interaction between the teacher and students (see next point), or can be recorded with or without supporting multimedia contents and uploaded to a website. Students can download or access the contents whenever they want afterward.
- *Active or passive*: Lectures can be delivered with or without an active interaction between the teacher and students, with an interaction that could be based on conventional communication strategies (question/answer approach) or multimedially enhanced (students can follow the lecture also visualizing multimedia contents or exchanging multimedia contents in real time or can interact with the systems for an interactive experience). Feedback can or cannot be given in both cases (i.e., open forum among students of the same course and/or teachers moderating the discussion, pre- and posttest to state the student learning level, exams, final dissertations) with or without an instant response (from a system or from a human moderator).
- *With or without the use of specific devices*: The courses can be delivered using conventional devices as laptop, webpages, and apps designed for smartphones or tablets or can use devices specifically designed for the aim of the course itself (see Sect. 10.2). This second approach can be of some interest for the delivery of courses for people working with children.
- *Interaction*: This can be a classic connection (Web connection + webcam and microphone) or enhanced (virtual gloves, virtual glasses as, for instance, in virtual reality settings [5] – a surgeon broadcasting live a surgical operation to an audience of students that could see the operation fields through the operator's eyes).

10.2 Teaching and Assessing Children's Needs

All those approaches have to be considered, keeping in mind the idea that the one who is going to exploit this delivered knowledge will be a child. This is crucial especially when the aim is to simulate an interaction with young patients or a young user. Role-play games and simulations, delivered through ICT tool, are a possible key to learn how to deal with children. The use of multimedia and online resources can help educators and students to focus on the main aim of the education process that in this case is not only the transfer of knowledge but also the acquisition of specific skills.

A specifically designed environment can help the learner to understand the children's needs, while offering real-time feedback on the advancements of his or her training. As long as this includes the use of words and examples that the child could understand, a social Web simulator can be created, in which the adults can level their communication strategies to the ones of the virtual children they have to communicate with.

Besides, the early detection of distress, discomfort, or sufferance can also be taught. The main indicators of cyberbullying, Web-based harassment, or Internet exploitation can be showed to students in simulated scenarios that make use of the same technologies in which those assaults take place. Besides, the early detection of psychological discomfort, also secondary to a child's illness, can be retrieved starting from the analysis of the child's social life and interactions with the real and the virtual world (see also Chap. 6). The strategies aimed to reach this goal have to be learned on the Internet or on the communication device that children use every day and where most of these situations happen.

In addition, most of the clinical procedures that health workers have to perform on children are complicated. This is due to two main factors: the first one is the fact that children are small, and it means that operating on a child is more difficult than operating on an average adult. The second reason is that children do not like to be harmed. This is a key issue, and there is no point in learning a procedure if the child does not allow you to perform it on him or her. The construction of settings in which a student can simulate an interaction with a child, also in the case of a invasive or noninvasive procedure, can be achieved with the use of specific tools: the ones that could simulate the child's body plus the interactive devices – also based on online resources and database – that could help to recreate the real scenario in which the health workers will have to operate.

In distant learning, the use of virtual reality can help to reach this goal, thanks to the use of more basic devices that could allow the student to work on virtual bodies. This can be crucial especially in remote or rural areas where teaching centers with expensive simulation room cannot be built.

Conclusions

Working with children means the achievement of peculiar social and communication skills. It also implies a number of issues – as children's safety and child abuse and harassment prevention – that are part of the common practice today for those that live and work in strict contact with the youngsters.

That knowledge can be taught. Especially in remote areas, where the availability of trained teachers is poor, the use of eLearning and distant learning approaches can offer a valid solution, able also to exploit the potentialities of the most advanced technologies. Simulation can be created in order to make the learning process easy and to prepare adequately the students for those challenges they would experience once they work in real settings.

The goal is to reach a great number of students, with accessible and sustainable tools, in spite of the physical location they live in, offering an environment in which they could test themselves and learn to think as a child does. It implies not a top-to-bottom communication with the child but a side-by-side interaction, in which the smallest one could easily express themselves and find a comfortable way to create relationships.

The final aim, in fact, is the well-being of the children, and their central role, especially in the management of their own health, has always to be considered, above all by those that will be the ones that would work for that.

References

1. Jean P (1997) La construction du reel chez l'enfant, 6th edn. Delachaux et Niestlé, Neuchatel/Paris. ISBN 2-603-00741-6
2. Schor EL (1999) Caring for your school-age child: ages 5 to 12. American Academy of Pediatrics, Bantam. ISBN 0553379925
3. Lau RR, Williams HS, Williams LC, Ware JE Jr, Brook RH (1982) Psychosocial problems in chronically ill children. J Community Health 7(4):250–261
4. Allen IE, Seaman J (2014) Grade change: tracking online education in the United States. Babson Survey Research Group. Higher Education Reports. ISBN 9780984028849 electronic resource, available from: http://babson.qualtrics.com/SE/?SID=SV_7R2QI3e65TBexXn
5. Cemenasco AF, Bianchi CC, Tornincasa S, Bianchi SD (2004) The WEBD project: a research of new methodologies for a distant-learning 3D system prototype. Dentomaxillofac Radiol 33(6):403–408

Health eDucation: Teaching Healthy Lifestyles for a High Quality of Life

11

Fabio Capello

The Internet is nowadays the most accessible source of information in every field of the human knowledge. Yet it is common experience that not every content found on the Internet can be considered reliable. On the contrary, much false or misleading information are published and shared every day in forums, webpages, or social networks. In addition, everyone develops his or her own opinion, so that also for given or scientific facts, the reports are often conflicting among themselves. This is a critical issue that is apt to lead the world to a Tower of Babel, in which every single matter can raise objections: no matter how reliable or unreliable a source of information can be, a choir of voices that support or dissent it would always be present.

On the other hand, the Internet – together with the communication technologies and devices that are going to be developed in the near or in a far future – can offer a valid support for the delivery of that knowledge that could help families and children to achieve a better health and therefore an improved quality of life.

Talking about children, though, the first step is to consider the final audience the information is intended for. There is no point in creating an information device that can be accessed only by adults, excluding in fact the real target of health education intended for children.

The wide spread of easily accessible and cheap smart-devices also among children can help to reach this goal. The next generation is likely to grow surrounded by touch screens and always-connected apparatuses. The extent of that is hardly foreseeable, today, but it is strongly thinkable that high-technological devices are only beginning to show their potentials.

Besides, children have their own way of communicating and they live in their own world, where many of the barriers or the constraints that limit grown-ups have not been built yet. This is an extraordinary resource that has to be exploited,

F. Capello, MD, MSc
Pediatrics and Child Malnutrition, CUAMM – Doctors with Africa,
Via S. Francesco, Padova, Italy
e-mail: info@fabiocapello.net

F. Capello et al. (eds.), *Telemedicine for Children's Health*, TELe-Health,
DOI 10.1007/978-3-319-06489-5_11, © Springer International Publishing Switzerland 2014

provided that the same health education (which those devices can help to achieve) has to help children deal with those tools, taking advantage from them without becoming alienated.

11.1 The Use of Communication Devices to Deliver Health Education Among Children

In science communication, the golden standard should be offering to an audience the most reliable information (so that the contents that are broadcast are not distorted when compared to the original ones) in the most understandable way. It comes to itself that the more the information is simplified and the plainer the language used (and therefore more understandable by a nonspecifically educated audience), the more the accuracy of the message – when compared to the proper scientific information – is compromised and vice versa. The theory of relativity, for instance, or the Higgs' boson can be explained through examples that make use of people's everyday experiences (a train in motion for the Einstein's theory or a field of snow for the Higgs). Nonetheless, the examples given are not as accurate as the mathematic formulae that lie behind. But how many people are able to understand advanced math and therefore get the essence of the scientific message?

Health-related contents should not be distorted, but how can a complex scientific information be delivered to the public so that it can be easily understood? How can it be possible to give the same information to an audience of children? Besides, giving information, delivering the message, does not mean to promote and acquire a healthy behavior (every smoker knows that smoke is potentially lethal; nonetheless, that information does not prevent them from smoking cigarettes).

Those are questions that every professional that tries to implement health educational projects has to deal with. The use of Internet-based strategies, yet, can help to overcome some of those issues, provided that developers keep clear in their minds that the final audience are the families, and above all the same children those messages are intended for.

A starting point has to be the differences that distinguish children from adults: what is inconvenient for grown-ups may not be for kids. Fantasy, creativity, and imagination are strong instruments on which can be build upon. Those are key issues that have to be strongly stressed in a child's everyday life. But they can also become useful means in the development of health educational programs for children that made use of ICT tools. Friendly online interface shown on portable screen can be used, for instance, while the child lied sprawled on the floor or on a bed (Fig. 11.1). Is this child just lying on his couch, messing around with his tablet, or is he traveling on a spaceship toward outer planets at the speed of imagination?

Health education is not a game, but the way it can be delivered can be. Education in fact means not only to learn some concepts but also to interiorize them, so that the information acquired can become future behaviors. On the other side, children cannot be left alone in this process.

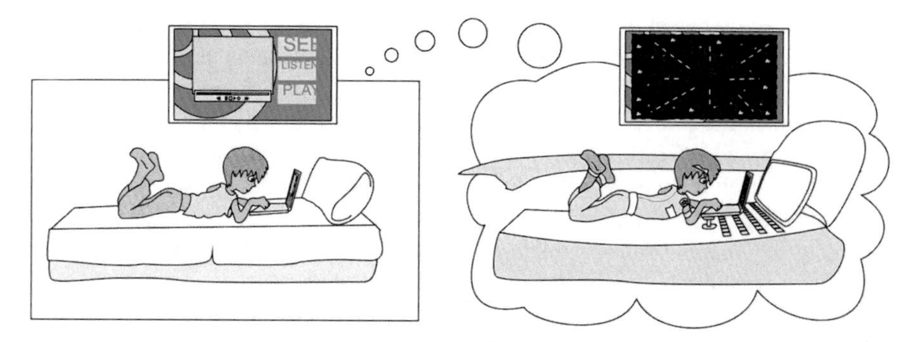

Fig. 11.1 The use of kid-friendly devices can exploit the natural curiosity of the children. Creating a fertile environment where information can be shared and capacity can be acquired is crucial. Yet children have to become part of the world we would not bring them into. Educational programs have to consider therefore the natural aptitudes of children, together with the richness of their inner world that is often made of creativity and imagination. Those talents have to be fostered and enhanced, so that the educational process can become an active one, where the child can develop a deep awareness of the problem, setting it inside an inner web of positive constructions. Being a space captain that travels around the universe to stop the evil germs can be an easier way to teach children to wash their hands, rather than the mere information that dirty hands carry harmful bacteria that can make their tummies hurt

11.2 How Can Health Reach Children in an Always-Connected Era?

As we have seen, ICT tools are powerful means to deliver education to children. Yet which strategies are most convenient in order to deliver the correct message in the most profitable way?

The use of video games can be an interesting field of application. Exploiting something that they already love and know, children can acquire awareness of health problems that could last over time [1]. Children can be amused by the game and educated by the underneath message.

Although there are still some controversies about its real usefulness, this strategy has been already proved for health professionals [2] with interesting results.

> Simple society games, which are accessible and enjoyable to players of all ages, and online worlds, which offer a unique opportunity for narrative content and immersive remote interaction with therapists and fellow patients. Both genres might be used for assessment and training purposes, and provide an unlimited platform for social interaction. The mental health community can benefit from more collaborative efforts between therapists and engineers, making such innovations a reality [3].

Creating a group of interest can also help to spread the knowledge among children and families. Some experiences show how contents created by families and patients of a same health environment (as children treated in the same hospital) can help to promote healthy behaviors or trustworthy information related to health, well-being, and treatments [4]. Virtual magazines, as well as forum or social network, based on the production of contents by the same families and by the professionals that treat those same children every day – contents easily accessible as they are online and just a click away – can offer precious resources to other families: for those who want to find information related to health issues (promotion of healthy lifestyles, reduction of potentially harmful behaviors) and for those that already have a child with a health problem at home and would like to receive consistent information related to a particular condition, procedures, or prevention strategies.

Web 2.0 offers an interesting solution, as long as those contents that are due to spread health education can be mediated and rated, so that only those resources that can be of some interest and above all that are reliable can effectively reach those families that would like to promote healthy lifestyle for their offspring and boost the quality of life for their children.

Some tools have already proved to be useful in the delivering of information related to healthy lifestyles. The intake of junk food as well as sedentary behavior is an emerging issue especially in Western countries in which children are even more apt to spend their time in indoor activities and inside strict routine, where very few space is left for those old-fashioned games children all around the world have always been through.

Involving families and children in this process has been proved to give good results [5]. Paradoxically, the use of screen-based devices that could help families with distant support and information can help to reduce the time children spend in front of a screen (playing video games, watching TV, instant messaging with friends), promoting healthy behaviors or helping families to schedule healthier diets or physical activities for their children.

11.2.1 eDucating in the Rural Contexts

A special consideration is due in relation to those settings in which the delivery of health education and of health information can be difficult, both because of cultural restraints and of physical or geographical limitations. Promoting health educational campaigns in rural districts could be tricky: in those countries in which children can attend schools also in very rural areas, educational programs can be delivered in face-to-face encounters; on the other side, in many developing countries, children are likely to have an easier access to mobile devices and Web-based resources than frontal lessons.

Mass media can play a role, with top-to-bottom campaigns that could help to spread the knowledge. In many countries, radio and TV have already a strong coverage, and the population is keen to receive and adopt those messages delivered through those media. On the other hand, mass communication projects have been proved to be lesser effective in the reception of messages that are due to change habits and behaviors in the population. Tailored information campaigns are needed, although the limitations related to rural, poor, and low-educational settings are well known.

Distant learning has been already experimented with different strategies for both children and professionals with discordant outcomes [6–8]. The use of ICT tools for the delivery of health education-related information, though, can set a different goal, thanks to the use of new strategies that take also in account the peculiarities of the single user to whom the information is intended for, children with impairments included [9, 10]. Facebook and Google Search, for instance, already make use of algorithms that tailored the ads according to the search queries of the activities of a user on the network. Push-and-pull information can be sent to children browsing the Internet, so that appropriate contents can be acquired. The use of mobile technologies can augment the extents of these programs, with apps that can follow the child everywhere (smart-devices' sensors can already monitor the everyday activity of a child, in spite of the geographical displacement of the child itself), guiding the kid in its everyday life, promoting and enhancing the adoption of healthy lifestyles. This is due to increase the chance that the correct message can be delivered to the correct child in the most appropriate way.

11.2.2 eDependence or eDucation?

The Internet and the connecting devices can therefore help children to achieve a better health. They can learn what can be healthy and what can be safe, and the use of electronic tools can help to develop permanent imprinting that can deeply affect the behavioral attitude of the child. The end of it is supposed to be the adoption in a natural way and with no conflicts of healthy behaviors and lifestyles.

Yet because the risk is to develop a constant addiction to whatever can be plugged, dissolving the natural ability of children to design their own world around them, professionals have to be involved.

This is the way an educational approach, also based on ICT tools, should take place where education is at home, namely, in the schools. Teachers can support children when a health problem arises, when a prompt intervention is needed, or when unhealthy behaviors have to be prevented. Computer-based questionnaires (also needed to assess the psychological status or discomfort of a child [11]), role-play games, and multimedia can be used at school and – when needed – can continue at home.

School-based programs that make use of the communication technologies and of those easily accessible devices children already master have to be fostered. Children in fact cannot be interested in browsing health-related contents on their own. On the contrary, when properly driven, they could find the world of health education fascinating. Once encouraged to understand the problems that make them sick from time to time, or those illnesses or discomforts they suffer from, but that they are too shy to disclose to the others (friends, teachers, doctors, and often families), they can become more constructive and more interested in the management of their own health.

It comes by itself, though, that it can be a very tricky pathway to follow: pushing children to use computer and smart-devices can restrain them from those same threats we would like to avoid. This is why, even if is paramount that a tool for health education has to be designed so that a child can use and understand it by itself, parents, teachers, educators, and health professionals have to be constantly around, supporting the child and helping him to switch off.

Conclusions

Telehealth means also the delivery of useful information that could contribute to educate children to health. It means the promotion of healthy lifestyles and the prevention of harmful behaviors, thanks to the use of Web-based technologies or of communication devices.

As long as children are already eager to use something they know, understand, and sometimes love, the communication channel to deliver those messages is already open. This is indeed a point of a wedge that can open a breech. Once the children are actively involved in educational programs that make use of communication technologies, the establishment of positive and healthy behaviors can easily be fostered by parents, teachers, and health professionals.

These are due to produce an imprinting in the child that is apt to guide him or her in the road to become an adult and possibly a healthy one.

References

1. Williams O, Hecht MF, Desorbo AL, Huq S, Noble JM (2014) Effect of a novel video game on stroke knowledge of 9- to 10-year-old, low-income children. Stroke. 45(3):889–92.
2. Akl EA, Kairouz VF, Sackett KM, Erdley WS, Mustafa RA, Fiander M, Gabriel C, Schünemann H (2013) Educational games for health professionals. Cochrane Database Syst Rev (3):CD006411
3. Wilkinson N, Ang RP, Goh DH (2008) Online video game therapy for mental health concerns: a review. Int J Soc Psychiatry 54(4):370–382
4. Luca PD, Chan M, Basak S, Segal AO, Porepa M, Pinard M, Au H, Birken CS (2013) A qualitative description of the development and evaluation of our voice, a health promotion magazine created by pediatric patients for hospitalized pediatric patients. Hosp Pediatr 3(1):59–64
5. Fulkerson JA, Neumark-Sztainer D, Story M, Gurvich O, Kubik MY, Garwick A, Dudovitz B (2014) The healthy home offerings via the mealtime environment (HOME) plus study: design and methods. Contemp Clin Trials. doi:10.1016/j.cct.2014.01.006, pii: S1551-7144(14)00016-0
6. Chiou-Fen Lin, Meei-Ling Shyu, Meei-Shiow Lu, Chung-I Huang (2012) Examining the effectiveness of a distant learning education program: use of patient safety and reporting law and ethics as example. Lect Notes Comput Sci 7513:381–385
7. Lih-Juan ChanLin, Hong-Yen Lin, Tze-Han Lu (2012) College's students' service learning experience from e-tutoring children in remote areas. Soc Behav Sci 46:450–456
8. Priyaa S, Murthy S, Sharan S, Mohan K, Joshi A (2013) A pilot study to assess perceptions of using SMS as a medium for health information in a rural setting. Technol Health Care. 1;22(1):1–11.
9. McCarthy M (2010) Telehealth or Tele-education? Providing intensive, ongoing therapy to remote communities. Stud Health Technol Inform 161:104–111
10. Feil EG, Baggett KM, Davis B, Sheeber L, Landry S, Carta JJ, Buzhardt J (2008) Expanding the reach of preventive interventions: development of an Internet-based training for parents of infants. Child Maltreat 13(4):334–346
11. Rooney RM, Morrison D, Hassan S, Kane R, Roberts C, Mancini V (2013) Prevention of internalizing disorders in 9–10 year old children: efficacy of the Aussie Optimism Positive Thinking Skills Program at 30-month follow-up. Front Psychol 4:988

Conclusions 12

Fabio Capello, Andrea E. Naimoli, and Giuseppe Pili

Children have been always considered the most valuable part of our society, in spite of the place or the time they lived in. Although their role changed a lot among the historical periods, they always remained the weakest point and at the same time the strongest resource for the human race. This is why their health intended as their well-being has to be fostered and nourished.

The quality of care, and therefore of life for children and families, has to be considered a universal value. Nevertheless, many are still the constraints that impede children to properly access health systems worldwide. Also in those countries where the standard level of care is very high, the burden associated with acute and chronic disease in children can be often unbearable.

On the other hand, telecommunication devices are today widespread also in the least developed counties, while the Internet and the Web-based application are becoming even more accessible to a greater number of people. Children of today and tomorrow are apt to naturally interact with those instruments that are already part of their everyday life (the so-called digital natives).

The use of ICT in order to establish a model that could deliver healthcare linking people from different places among themselves is a practicable and sustainable solution, those connections ranging from next-door consultations up to intercontinental referrals. The aim is to create a net of people affected by medical conditions

F. Capello, MD, MSc (✉)
Pediatrics and Child Malnutrition, CUAMM – Doctors with Africa,
Via S. Francesco, Padova, Italy
e-mail: info@fabiocapello.net

A.E. Naimoli
Tech Department, Airpim Inc., Wilmington, DE, USA
e-mail: andrea.naimoli@elementica.com

G. Pili
Department of Child and Adolescent Psychiatry and Neurological Disorders,
ASL 1 Imperiese, Consultorio di Sanremo, Sanremo (IM), Italy
e-mail: giuspili@gmail.com

F. Capello et al. (eds.), *Telemedicine for Children's Health*, TELe-Health,
DOI 10.1007/978-3-319-06489-5_12, © Springer International Publishing Switzerland 2014

or healthy people whose well-being is aimed to be preserved (children and families); of those that care for the latter (relatives, friends, schoolmates, teachers, coach, educators); and of professionals (family doctors, specialists, health workers, social workers) aimed to work together for the well-being of the youngest ones.

Yet a working model has to consider first the people those tools are intended for, which in other words means that the delivery of care has to be children centered and family oriented. The child has to become the owner of his or her own health and the main actor of his or her well-being, with the support of all the people that take care of him or her. It implies that systems have to be built so that even the child can understand them and possibly use them. Children of different ages and coming from different backgrounds should be able to access those applications that consequently have to adapt themselves according to the final user.

It also means that different levels of access have to be guaranteed so that professionals and caregivers could access the needed information with different grades of liabilities and reaching only those data they need to process. This is in order to protect the privacy and to avoid the handling of a complex set of unneeded data.

Starting from the family and the community, the natural environment of a child, telemedicine has to create that safe landscape that could recreate the once upon a time neighborhood, in which each one cares for the other children. In other words, a surveillance able to early detect discomforts of signals of distress that could mirror a more complex physical or psychological disease. Also the same problems related to the use of Web-based devices – as Web addiction or cyberbullying – should be watched over in a such designed scenario.

The health professionals are called to play a major role, assisting children, families, and communities and exploiting the potentiality that comes from ICT to increase the level of accuracy and effectiveness for prevention, diagnosis, treatment, and follow-up. A net of specialists, able to interact among themselves in spite of the distances – in terms of culture, space and time – that separate them, can be created, with the aim to bring the excellency also in those places that are normally cut off from access to the major health centers.

Those same specialists have to foster the home-based care, able to help children to feel more self-confident and less victim of the stress that surrounds any acute or chronic medical condition. The reduction of the burden and the improvement of the quality of life – as well as the diagnosis and the treatment of a disease – being always the final goal of any medical action.

This is due to create a protective environment in which children could feel safer and where those distances that separate ill children with the healthy ones could be filled.

However, health cannot be considered only a resource for those that could afford the expenditure of the care or for those that are lucky enough to live in a developed area. Besides, every place on Earth could become in a matter of hours affected by those same misfortunes that affect very poor regions. For those rural areas and in extreme scenarios as the ones that follow natural catastrophes, a war, or a major accident, telemedicine could offer a valid support, also in consideration that children are the ones most at stake in these situations.

In any case, to reach this goal, it is crucial to choose wisely among the possible solutions available today and in the future, also considering those scenarios that are still way to become real. It also means that those resources that are already easily accessible and that exploit the existing infrastructures have to be considered. Devices designed for a specific purpose can be implemented in major health centers as well as point-of-care devices can be installed in the minor health units in order to increase the net of assistance, offering the same condition of healthcare to different locations. Smart-devices, personal computer, and mobile technologies instead have to be used to reach, in a widespread distribution, children and families in their home or in the places they normally live.

This is a possible way to deliver health, to promote prevention and healthy lifestyles, and to educate children, families, and professionals to better care for the most vulnerable ones, reducing the distances and improving the quality of life: for children today, for happy adults tomorrow.

Further Reading

1. UN General Assembly (2000) United Nations Millennium Declaration. Resolution 55/2. 8th plenary meeting, New York. 8 Sept 2000
2. UNICEF (2004) The state of the world's children 2005. The United Nations Children's Fund, New York. ISBN 92-806-3817-3
3. WHO – World Health Organization (2011) Telemedicine. Opportunities and developments in Member States. Report on the second global survey on eHealth. Global observatory for eHealth series, vol 2. New York. ISBN 978-92-4-156414-4 http://www.who.int/goe/publications/ehealth_series_vol2/en/